PRACTICAL SPECTROSCOPY

A SERIES

Edited by Edward G. Brame, Jr.
Elastomer Chemicals Department
Experimental Station
E. I. du Pont de Nemours and Co., Inc.
Wilmington, Delaware

Volume 1: Infrared and Raman Spectroscopy (in three parts)
edited by Edward G. Brame, Jr. and Jeanette G. Grasselli

Volume 2: X-Ray Spectrometry
edited by H. K. Herglotz and L. S. Birks

Volume 3: Mass Spectrometry (in two parts), Part A
edited by Charles Merritt, Jr. and Charles N. McEwen

ADDITIONAL VOLUMES IN PREPARATION

Mass Spectrometry

(in two parts) **Part A**

Mass Spectrometry

(in two parts) **Part A**

edited by Charles Merritt, Jr.

Food Sciences Laboratory
U.S. Army Natick Research and Development Command
U.S. Department of Army
Natick, Massachusetts

Charles N. McEwen

Central Research and Development
Department
Experimental Station
E. I. du Pont de Nemours and Co., Inc.
Wilmington, Delaware

MARCEL DEKKER, INC. New York and Basel

Library of Congress Cataloging in Publication Data

Main entry under title:

Mass spectrometry.

(Practical spectroscopy ; v. 3)
Includes indexes.
1. Mass spectrometry. I. Merritt, Charles [Date] II. McEwen, Charles N. [Date] III. Series.
QD96.M65M37 543'.087 78-24085
ISBN 0-8247-6749-7

MARCEL DEKKER, INC.
270 Madison Avenue, New York, New York 10016

Current printing (last digit):
10 9 8 7 6 5 4 3 2 1

PRINTED IN THE UNITED STATES OF AMERICA

PREFACE

This book is Part A of the Mass Spectrometry volume of Practical Spectroscopy and, like Part B to be published subsequently, is devoted entirely to the field of mass spectrometry. Other volumes in the Practical Spectroscopy Series will cover infrared and Raman spectroscopy, x-ray spectroscopy, nuclear magnetic resonance spectroscopy, ultraviolet-visible spectroscopy, emission spectroscopy, and so forth.

The aim and objective of this volume is to cover a variety of diverse areas where mass spectrometry is of practical value. Topics include application of mass spectrometry to protein structure determination, to the study of drug metabolism, to the food and flavor industry, to chemical derivatization, to encapsulated environments, etc. Through these examples of the practical applications of mass spectrometry it is hoped will come a better understanding of the scope of this technique and a cross-fertilization of ideas and methods among the various areas of application.

Because mass spectrometry is a rapidly developing field, there are several relatively new areas such as field desorption, collisional activation analysis, and negative-ion chemical ionization which are unlikely at this time to have been used extensively in practical areas of application. We have thus included chapters concerning the analytical applications of each of these methods.

Volume 3, Part A contains five chapters. The first, appropriately, is an introductory chapter authored by Professor Herbert Budzikiewicz detailing the history of ion beam chemistry. Chapter 2 deals with the application of mass spectrometry in protein structure determination and is authored by Professor Robert E. Lovins. Chapters 3 and 4 describe the application of mass spectrometry to specific areas. Chapter 3, authored by Mr. M. Kolor, discusses the use of mass spectrometry in the food and flavors industry. Chapter 4, by Drs. F. E. Saalfeld and R. A. Saunders, reports on the analysis of encapsulated environments, in particular the analysis of submarine atmospheres. Chapter 5 covers analytical applications of field ionization and field desorption mass spectrometry and is authored by Professors Hans D. Beckey and Hans-Rolf Schulten. This chapter includes a brief section on the theory of these techniques, a section on experimental techniques, and an extensive section on applications of these techniques.

The editors are indebted to Dr. Thomas L. Pugh who participated as a co-editor in the early preparation of this volume. He was regrettably unable to continue, but his contributions to the concepts and organization are manifold and are gratefully acknowledged. We particularly appreciate his diligence in soliciting some of the manuscripts which appear here and which have greatly enriched this work.

Charles Merritt, Jr.
Charles N. McEwen

CONTRIBUTORS TO PART A

HANS D. BECKEY, Institut für Physikalische Chemie der Universität Bonn, Bonn, Federal Republic of Germany

HERBERT BUDZIKIEWICZ, Institut für Organische Chemie der Universität zu Köln, Cologne, Federal Republic of Germany

ROBERT E. LOVINS,* Department of Basic and Clinical Immunology and Microbiology, Medical University of South Carolina, Charleston, South Carolina

MICHAEL G. KOLOR, Central Research Laboratories, General Foods Technical Center, Tarrytown, New York

F. E. SAALFELD, Physical Chemistry Branch, Chemistry Division, Naval Research Laboratory, Washington, D. C.

R. A. SAUNDERS, Physical Chemistry Branch, Chemistry Division, Naval Research Laboratory, Washington, D. C.

HANS-ROLF SCHULTEN, Institut für Physikalische Chemie der Universität Bonn, Bonn, Federal Republic of Germany

---

*Current Affiliation: Heart Research Institute, Institutes of Medical Sciences, San Francisco, California

CONTENTS

CONTENTS OF PART B

Mass Spectrometry

(in two parts) **Part A**

Chapter 1

HISTORY OF ION-BEAM CHEMISTRY

HERBERT BUDZIKIEWICZ

Institut für Organische Chemie
der Universität zu Köln
Cologne, Federal Republic of Germany

HISTORICAL INTRODUCTION TO ION-BEAM CHEMISTRY

The oldest part of the "black art" of chemistry is expressed by its Dutch name *scheikunde*, i.e., the art of isolating a component from a mixture, in the widest sense. The separation of charged particles by their mass has its origin [1] in the work on canal rays, which were discovered in 1886 by Goldstein and were shown by Wiechert and Wien to be deflected by electric and magnetic fields in relation to their mass. In 1897 Thomson constructed a spectrograph in which an ion beam could be deflected by electric and magnetic fields: ions of different mass traveling on a parabolic path were recorded on a photographic plate. With the aid of such an instrument in 1910, the existence of the isotopes ^{20}Ne and ^{22}Ne was demonstrated, and with an improved version, ion abundances could be measured with the help of an electrometer. After World War I, both Dempster and Aston introduced mass spectrometers with focusing properties for ions of equal mass, which resulted both in a better resolution and in a higher sensitivity. While Dempster resorted to first-order direction focusing of a 180° deflection in a magnetic field, Aston used an electric sector.

The idea of combining both direction and velocity focusing in one instrument was pronounced in 1929 by Dempster, resulting in the double-focusing mass spectrometers in use today. They are essentially based on the geometries proposed by Herzog and Mattauch and by Nier and Johnson. Both these geometries have their advantages and their disadvantages. One has to weigh the importance of being able to use a photoplate for recording versus the value of having a focal point between the two sectors, which, for example, allows arrangement of the electric and magnetic sectors in either sequence. We owe the top performance instruments available today to the theoretical work of Mattauch, Ewald, and Hintenberger. Other brilliant ideas for ion separation, for example, the use of trochoidal and cycloidal ion paths as first suggested by Bleakney and Hipple in 1938, did not have any major impact on modern commercial instrumentation.

The separation of moving ions by magnetic and electric fields is still the dominant technique and thus far irreplaceable for applications such as exact mass measurements or investigation of metastable ions. Yet, there are other possibilities. In 1948 Stephens suggested a separation of ions of different mass by their different traveling speeds in a linear drift tube--the time-of-flight principle in use today for studying fast reactions. The so-called mass filters (the best known ones are the quadrupole instruments) go back to an idea of Paul and Steinwedel in 1953. Originally these mass filters were in use primarily for low-mass gas analyses; however, their resolution and sensitivity have been improved recently in such a way that they can compete successfully with magnetic instruments, especially for gas-chromatographic and chemical-ionization work. And there are many other systems for special applications, for example, the ion-cyclotron-resonance instrument based on an idea of Hipple that is used for studies of ion-molecule reactions.

The desire for the new, one of the strongest stimuli of mankind, resulted in another branch of chemistry: synthesis, the art of preparing something that was not available before. Thus, the next part of this historical review will be devoted to the formation of ions, and to start with, their formation from molecules in the gas phase. It takes energy to remove at least one electron from an atom or molecule to form an ion (between 4 eV for Cs and 25 eV for He, and around 10 eV for most organic compounds). The oldest and most widely used method (where, again, Dempster and Nier were pioneers) is to subject a gas stream to an electron beam, the mean energy of which has been standardized at 70 eV for many years, since a flat maximum is observed at this value in the ionization efficiency curves of most organic compounds. Electrons emerging from a hot wire and accelerated by some potential are not monoenergetic. As long as the electron energy distribution was reproducible, it also seemed not to matter that fairly large amounts of energy were transferred (up to about 15 eV beyond the ionization potential). Hence commercial instruments were equipped with electron-bombardment ion sources constructed for operation at 70 eV, and up to 1965 essentially all mass spectra of organic compounds were recorded in this way. Thus several important features of fragment formation were simply overlooked. Special ion sources, such as those constructed by Fox et al. [2] for producing defined low energies, were in use only for ionization potential measurements; otherwise, low energies were used mainly to suppress fragmentation. It was mainly the work of Spiteller-Friedmann and Spiteller [3] that demonstrated that certain specific fragmentation reactions could be observed only at low energies and that stimulated interest in more systematic studies at different ionizing energies. Another way to transfer energy is photon ionization, based on experiments by Terenin and Popov [4] in 1932, but first used in mass spectrometry by Lossing and Tanaka [5] in 1956. However, with the developing laser techniques, this method certainly will gain in importance. Removal of an electron from a neutral species by strong inhomogeneous fields is also possible, owing to a tunneling effect. This is the basis of field ionization, the development of which in the early 1960s is mainly due to the work of Beckey [6]. The transfer of minimal amounts of excess energy during field ionization allows the analysis of rather labile substances. Although based on a series of earlier experiments with ion-molecule reactions [7], for example, by Lind and Talrose, it was the work of Field and Munson [8] that introduced the technique of chemical ionization at about this time: Collisions of neutral molecules with ions result either in charge transfer or ion-molecule reactions. The opportunity of not having to rely on what an excited ion might do, but rather to perform ion chemistry, opens a wide field of application for chemical ionization in structure analysis and offers some opportunity for further research. The transformation of kinetic into excitation energy by collision of ions with neutral molecules has been used increasingly during the last few years in collision activation work to induce decomposition of ions in the field-free regions of mass spectrometers. Basic ideas are due to Jennings and McLafferty [9].

Negative ions can be formed by electron impact with low-energy electrons and by charge transfer [10]. In this context the work by Ardenne [11] should be mentioned, who in 1957 developed a special instrument, which was capable of obtaining spectra of substances not otherwise amenable to mass spectrometry at that time. The formation of ions from substances present in the condensed phase is also possible. This is relatively easy for stable inorganic species; as early as 1906 Gehrke and Reichenheim [12] discovered that ions could be evaporated from a heated surface. This method of thermoionization is still in use. In 1935 Dempster [13] suggested the use of spark sources for inorganic work. For organic compounds the field desorption technique is important. In the mid-1950s Müller discovered the desorption of ions from charged tips, and it was again the work of Beckey and his group [6] that developed this technique in the late 1960s. Highly labile substances are amenable to mass-spectrometric investigation by this method. New techniques for ion formation are reported in increasing numbers, such as secondary-ion mass spectrometry (e.g., Benninghoven [14]) and the ^{252}Cf source (Macfarlane and Torgerson [15]).

Once a chemist has obtained a new substance, he wants to characterize it and to investigate its properties. The history of the theories that were put forward in connection with the fragmentation of organic ions reflects, in a way, the history of organic chemistry, including superstition, intuition, using the dicta of authorities instead of experimental proof, and severe controversies. At the beginning of this century Thomson observed that $COCl_2$ not only gave ions corresponding to the various isotopic M^+, but CO^+, Cl^+, and O^+ as well. Early hopes that structure elucidation was thus possible by some kind of puzzle game were quickly dimmed when it was discovered that--to take the classical example-- the mass spectrum of isobutane contained an ion corresponding to $C_2H_5^+$, a structural unit not present in the original molecule. Desperate ideas, such as "sudden death" or the "hydrogen soup," were pronounced that assumed a collapse of the original structure with subsequent recombination to give stable ions-- an idea which curiously enough has been revived recently in a modified form to account for the results of some low-energy rearrangements usually referred to as scrambling. To characterize a compound, ion intensities were listed and at best some wiggly lines were drawn across a molecule to indicate which part of it might have been lost. Some vague ideas that a minimum number of rearranged hydrogen atoms should accompany fragment formation were taken occasionally as guidelines in these formulations, as in the proposals (1956-1962) [16] for the ring D-cleavage of steroids, i.e., $[M - (42 + R)]^+$:

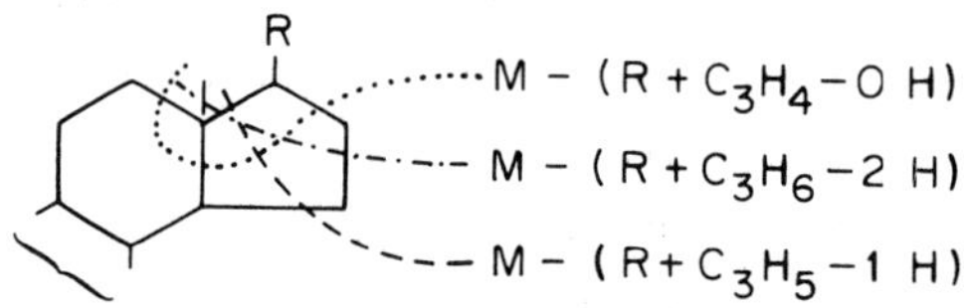

This idea was so deep-rooted that it was used as late as 1962 to assign a specific structure [17] to an alkaloid. Encouraged by observations of the behavior of compounds carrying functional groups, which apparently induced highly specific fragmentation reactions, a new period of mass spectrometry commenced after about 1950. Stenhagen, Beynon, McLafferty, Biemann, Djerassi--just to mention a few names-- are representatives of this era. The proverbial wiggly lines indicating where--presumably--an ion fragmented were replaced by fragmentation mechanisms essentially modeled after reactions in solution chemistry, taking into account especially ion and radical stabilities. It was the time of full arrows and fishhooks. This mechanistic approach to the interpretation of fragment formation hinges on the idea that a localized charge and/or radical site triggers reactions that either result in the pairing of unpaired electrons by bond cleavages or result in rearrangements that transfer a radical site to a better stabilized locus:

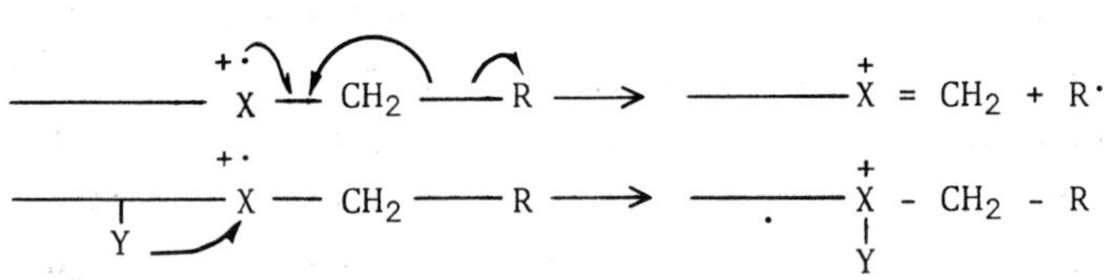

Certain processes were excluded, however, owing to some irrational taboos. Thus, skeletal rearrangements, i.e., the migration of atoms other than hydrogen, were not considered seriously [18] until after 1965. A considerable number of more or less pertinent experiments were devised to prove or disprove this concept and to explain exceptions, as demonstrated by emotionalized discussions concerning quasi-thermal reactions [19] around 1967.

The idea of the localized charge did not remain unchallenged, and indeed, it was highly vulnerable, especially early in its development. To combat the traditionalists, who denied any mechanistic relation between the structures of the un-ionized molecule and the fragment ions, fragmentation mechanisms had to be overemphasized. This resulted in the sometimes desperate attempt to explain mechanistically the genesis of any ion, which in turn entailed the odium that just post festum explanations were being offered. It cannot be denied that some misconceptions have been formulated by authors who were not too familiar with the basic principles of chemistry and physics. Thus, for example, Bentley and Johnstone [20] did not lack examples when they pointed out that vagueness and intuition were the main characteristics of the mechanistic approach. They were certainly right when attacking the abuses of the concept, but they were wrong as far as the basic idea was concerned, which has been proven to be sound: There are specific reactions occurring with high yield that can be correlated readily with the unaltered structure of the original molecule and that are linked to radical sites in the ion. But one should not forget that a mass spectrum is a product analysis of a system of competitive and consecutive reactions; thus the result is a function of time and energy. Once this has been realized, it becomes obvious why a correlation between structure and mass spectrum is not possible as it is for nuclear magnetic resonance.

Incidentally, Bentley and Johnstone were among the main advocates of an approach to the interpretation of mass spectra that may be called "energetic." It is obvious that if there are two competing reactions occurring with roughly identical rates, the one requiring less activation energy will occur preferentially. This has been demonstrated by Williams and Howe [21] for bisubstituted aromatics using $\Delta H°$ values and in a more empirical way by Remberg and Spiteller [22] for bifunctionalized aliphatics. There are several limitations to this approach that severely affect the step that goes from explaining to predicting fragmentation: The fragmentation behavior of the monosubstituted compound has to be known and there must be no interaction between different functional groups, as for example, has been shown to occur by Hesse [23] as S_Ni-type reactions. Of necessity, entropic factors that are of importance in rearrangement reactions are ignored, and, in order to use heats of formations, the structure of all particles involved has to be known. Yet in many instances the predictive value of the energetic approach cannot be denied, and, for example, both Stevenson's rule [24] and Audier's rule [25] are just special aspects of it. And, after all, the "energetic" and the "mechanistic" approaches are basically identical: The organic chemist has developed a good feeling for the energetic and spatial requirements for reactions to occur and thus in predicting which is most likely to occur. This explains why the mechanistic approach gives positive results in so many cases.

Completely different approaches for explaining mass spectra come from theoreticians. The best known among these theories is the so-called quasi-equilibrium theory [26] (QET) developed by Rosenstock, Wallenstein, Wahrhaftig, and Eyring in 1952 and subsequently modified several times. The basic idea is that if in a system of competing and consecutive reactions one knows all the rate constants, obviously one can calculate the abundance of any product after any given time of reaction. The problem is only to obtain reliable values for the various rate constants. The QET departs from the idea that prior to fragmentation ions exist in the lowest electronic state of excitation and their excess energy is quickly and freely fluctuating between the various accessible vibrational and rotational degrees of freedom. Bond rupture occurs whenever enough energy is accumulated in a given bond. Calculations proved extremely difficult, and a series of simplifications have been suggested; the last step being the well-known formula reduced to an entropic, an energetic, and a vibrational factor. Though early calculations seemed to be successful, e.g., for propane, later data from labeled compounds exposed the difficulties with rearrangement reactions: In order to calculate rate constants, the structure of an ion has to be known. A vicious circle develops if one does not try to explain the ion abundances in a spectrum, but rather tries to determine the structure of an unknown: to use the formulas provided by QET, one has to know the structure one actually is looking for. Nevertheless, many features of a mass spectrum can be explained, such as the kinetic shift (the influence of tight transition states upon the slope of an E vs k curve resulting in the preponderance of rearrangement reactions at low energies), the differences among mass spectra obtained from decays in the source and in the field-free regions, and so on [27]. Sidelines of the QET are the correlations of the so-called Z values with Hammett σ constants as attempted by Bursey and McLafferty [28] in 1966 and, more recently, their correlation with ionizing energy as proposed by Deroque [29].

At first glance the mechanistic approach discussed previously is in direct contrast to the main idea of QET: a triggering localized charge or radical site versus a free fluctuation of energy causing fragmentation. Many disagreements were along these lines. It was not until the late 1960s when Williams and Howe [27] tried to reconcile the contradicting ideas by pointing out that preferred fragmentation should occur from highly populated low-lying electronic states [30], which corresponded broadly to ions where an electron had been removed from a nonbonding pair or from a π

system. It is a synthesis of ideas derived from the kinetic approach of QET and from the mechanistic approach in combination with energy considerations that today gives the best understanding of what happens between ionization and the recording of a mass spectrum and that even allows to some extent a prediction of fragmentation patterns: thus, for instance, the relative importance of the various competing reactions known to be initiated by the various centers in a molecule at least can be estimated in many cases for low and high energies.

Molecular-orbital (MO) calculations have become more important recently, as, for example, the work of Lorquet [31] and Hirota [32]; but the use of many simplyfing assumptions, which are necessary to cope with the more complex molecules, may be the reason that no major breakthrough has been achieved. These difficulties are demonstrated by calculations by Dougherty [33] (perturbation and valence-electron MO calculations) and by Boer et al. [34] (self-consistent-field calculations) which arrive at contradictory results regarding the question of whether the McLafferty rearrangement of ketones is a concerted or a two-step process.

One fact remains: all theoretical calculations as well as all mechanistic formulations depend on the knowledge of ion structures. The auxiliary assumption that the structure of both the M^+ and fragment ions resembles essentially that of the un-ionized molecule is certainly a starting point, but not more than that. Therefore, the last part of this review will be devoted to another branch of chemistry, which concerns structure-elucidation methods. Structure-elucidation of ions produced in a mass spectrometer was in its earliest days a task perhaps not as hopeless as that of Sisyphus, but, as it turned out, frequently a fight against windmills. The first and most trivial question of any chemist which was about elemental composition, was a major problem. Natural isotope abundances were only of limited use. Beynon et al. [35]

could demonstrate in 1959 that elemental compositions could be obtained from high-resolution measurements, but it took quite some time before precision mass measurements became commonplace. In 1961 Biemann et al. [36] resorted to ^{15}N labeling in order to sort out the nitrogen-containing fragments from amino acids, and in an introductory book [37], which appeared in 1964, there are just 15 entries in the index under the heading "high resolution," more than half of which just point out where it would be desirable.

Isotope labeling in a systematic way was certainly a major step forward in ion structure elucidation. Here Djerassi and his group should be mentioned, from whose work stems a great deal of our knowledge about a number of classes of compounds. From labeling experiments we know that in aliphatic amines there is no hydrogen scrambling prior to α-cleavage and subsequent alkene elimination, even in the m* spectra; that during the McLafferty rearrangement a γ-H is transferred; that hydrogen scrambling is incomplete for the C_2H_2 loss from styrene after 10^{-6} sec, but complete after 10^{-5} sec, while in the $C_7H_7^+$ ion, randomization is complete before fragmentation occurs in the source. In principle it is possible to establish for any fragment ion which part of the original molecule it contains.

Heats of formation have been used to make probable or to exclude certain structures. Since the H° of acetone, propylene oxide, allyl alcohol, and methyl vinyl ether did not agree with that calculated for the $C_3H_6O^+$ ion stemming from a McLafferty rearrangement, this was taken as evidence for the enol structure of propene-2-ol [38]. Or, a sudden change in the appearance potential of the amine fragment for the butane derivative in the homologous series of mercapto amines was taken as an indication of an interaction of the two functional groups in a six-membered cyclic transition state [39]. There are obvious limitations: ΔH values of more complex species frequently can be obtained from increment calculations only, and may be rather similar for isomeric species; ionization and appearance-potential (AP) values are not always reliable and accurate enough; and the simple correlation that AP = $\Sigma\Delta H°$ (products) - $\Delta H°$ (educt) does not allow for activation energies and kinetic shifts. Thus, the problem of the structure of the $C_8H_8^+$ ions formed by the loss of C_2H_4 from tetralin in connection with the question as to whether this ion was formed by a retro Diels-Alder reaction could not be solved in this way [40].

The identity of organic compounds has frequently been established by comparing some of their respective properties (melting point, spectra, and so on). This approach has been used in mass spectrometry as well. By comparing the mass spectra of anthraquinone and fluorenone, Beynon suggested that the $[M - CO]^+$ ion of anthraquinone and the M^+ of fluroenone [41] had the same structure. From the comparison of the mass spectra of 3-methylcyclohexanone and menthone it was concluded that the ion formed by McLafferty rearrangement of the latter was not a ketone [42]. Biemann's shift rule is based on an extension of this method, which was first used in 1961 [43] to propose a structure of the alkaloid quebrachamine and which proved to be of great importance in the structure elucidation of many indole alkaloids. It implies that certain substituents will hardly alter the gross fragmentation behavior, and therefore, the mass spectra of related compounds will resemble each other, allowing for the mass increment of the substituent. It is obvious that there are limitations to the spectra comparison, which, incidentally, are inherent to all methods to be discussed subsequently. Isomeric ions may have mass spectra that are almost indistinguishable; ions of identical structure but of different origin may have different amounts of internal energy, as, for example M^+ of Ph-O-Ph and $[M - CO_2]^+$ from diphenyl carbonate [44], and thus will give more or less different ion abundances. And there is a rather important question: Is it possible to describe the sum of ions of a given elemental composition by a single structure--classical or nonclassical--even if obtained by a single ionization or decomposition process? From the early work of Grubb and Meyerson [45] we are acquainted with the problem of the structure of the C_7H_7 ion (benzyl or trophylium or a fluctuating equilibrium) and about the complete C and H scrambling prior to further fragmentation. It is a question of energy barriers and of the structural requirements for isomerization and decomposition, a question of the energy distribution of the ions.

Ions formed in a conventional electron-impact source have different amounts of energies, and owing to the high vacuum, there is obviously no equilibrium distribution of energy: Excess energy may range from amounts too low to allow any reaction at all up to highly excited species, which may undergo the most unlikely reactions. Furthermore, these ions may have any lifetime up to about 10^{-6} sec. To circumvent this difficulty research has been concentrated recently on ions that have fairly uniform internal energy and a lifetime of approximately 10^{-5} sec, viz. the metastable ions [46].

Metastable ions supposedly were observed first by Aston, and in the beginning they caused much confusion as the mass spectrum of H_2 showed an ion of mass 0.5. The conclusions that might have been drawn are obvious. A theoretical explanation was given by Hipple et al. [47] in 1947. As long as the main concern of mass spectrometry was to obtain highly reproducible mass spectra, bumps in the baseline were obviously evil, and older instruments had so-called metastable suppressors. When the interest in fragmentation processes was developing, the utility of m* was recognized since it correlated mother and daughter ions, but the idea that all atoms of the neutral lost in a metastable transition had to belong to a structural subunit of the decomposing ion had to be given up in 1966 [48]. IKE, MIKE, and DADI spectrometry demonstrate the increasing importance for structural work.

Papers on ion structures relying on metastable ion data have been filling the pages of the journals recently. The methods used comprise a comparison of the m* spectra obtained in the first or second field-free region to establish the identity of some ion with one (hopefully) of known structure [49]. What has been said about the comparison of 70 eV spectra is pertinent here, too, as evidenced by the observation that the m* spectra of $C_3H_6O^+$ ions obtained from single and double McLafferty rearrangement were not identical [50]--though, as shown later, their structures were. To circumvent this difficulty collision activation is used; this gives identical energy distributions because ions with almost no excitation energy are investigated. Modern research of metastable ions is intimately connected with the work of Beynon. He realized that the shape of metastable ion signals, sharp or flat-topped, depended on the energy liberated during fragmentation and that information about the fragmentation process and ion structures could thus be obtained [51]. Comparison of peak shapes is used increasingly today to establish the identity of ion structures.

Spectra comparison is certainly an important tool for the structure elucidation of ions--but data for comparison have to be available. Chemistry, in the sense of probing a structure by reactions characteristic for specific functional groups or structural units, has become possible with two other techniques. Ion cyclotron resonance allows selection of any long-lived ion and subjection of the trapped ion to ion-molecule reactions. One of the first examples of this technique was given by Dickman et al. [52], who established the enolic structure of ions derived from both the first and second McLafferty rearrangement. Chemical ionization--here the work of Hunt [53] should be mentioned--relies on ion-molecule reactions as well; but the importance of the method lies not so much in the structure elucidation of ions but rather in specific reactions of certain functional groups.

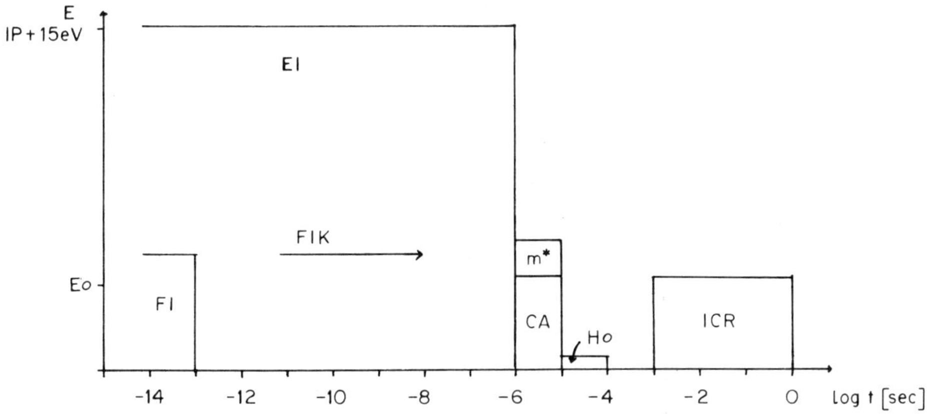

FIG. 1. Schematic diagram of portions of the time-energy continuum of ions covered by various mass-spectroscopic methods (the numbers have to be taken as a rough indication only). E_0 is the minimum activation energy for decomposition of a specific ion, hence for $E < E_0$ nondecomposing, but possibly rearranged ions, and for $E > E_0$ decomposing (nonrearranged or rearranged) ions are investigated. EI: electron impact; FI: field ionization; FIK: field-ion kinetics; m*: decomposition in the field-free regions; CA: collision-induced decomposition in the field-free regions; ICR: ion cyclotron resonance; H_0: thermodynamic data (appearance potential, heat of formation).

As was mentioned previously, the appearance of a mass spectrum is both a function of energy and of time. A comparison of mass spectra obtained from decomposition reactions in the ion source (up to $< 10^{-6}$ sec) and from the various field-free regions ($\sim 10^{-5}$ sec)--especially in context with labeling experiments--gave some indication about the rate of slow reactions. Fast reactions down to about 10^{-11} sec can be probed today by field-ion kinetics [6]. These measurements explain why a series of mass spectra obtained with a conventional electron-impact source are rather structure-unspecific. Rearrangement reactions with low activation energies and not too severe steric requirements had had ample time to proceed.

The armamentarium for obtaining structural information of ions formed in the mass spectrometer has been developed considerably in recent years, as well as an understanding of the complex system of competing and consecutive reactions, and their dependence on energy and time. From this it became obvious that the various methods are not necessarily measuring the same ionic species (Fig. 1), and results that seemed to be contradictory at first glance turned out rather to be complementary.

This short review could not give credit to everyone who contributed to knowledge in mass spectrometry. Especially in recent years many groups have been engaged in successful and important research. Only a rough cross section could be given here. No evaluation whatsoever is intended.

REFERENCES

1. For literature on the beginnings of mass spectrometry, see (a) J. H. Beynon, *Mass Spectrometry*, Elsevier, Amsterdam, 1960, Introduction and Chap. 1; (b) H. Kienitz, *Massenspecktrometrie*, Verlag Chemie, Weinheim, 1968, Chap. A 1; (c) J. H. Beynon, *Int. J. Mass Spectrom. Ion Phys.*, in. press.
2. R. E. Fox, W. M. Hickam, D. J. Grove, and T. Kjeldass, Jr., *Rev. Sci. Instrum.*, *26*, 1101 (1955).
3. For examples, M. Spiteller-Friedmann and G. Spiteller, *Chem. Ber.*, *100*, 79 (1967).
4. A. Terenin and B. Popov, *Phys. Z. Sowjetunion*, *2*, 299 (1932).
5. F. P. Lossing and I. Tanaka, *J. Chem. Phys.*, *25*, 1031 (1956).
6. For literature see H. D. Beckey, *Field Ionisation Mass Spectrometry*, Pergamon, New York, 1971.
7. For literature see C. E. Melton, in *Mass Spectrometry of Organic Ions* (F. W. McLafferty, ed.), Academic, New York, 1963, Chap. 2.
8. M. S. B. Munson and F. H. Field, *J. Am. Chem. Soc.*, *88*, 2621 (1966).
9. K. R. Jennings, *Int. J. Mass Spectrom. Ion Phys.*, *1*, 227 (1968); W. F. Haddon and F. W. McLafferty, *J. Am. Chem. Soc.*, *90*, 4745 (1968).
10. For a review on early work see Ref. 7, Chap. 4.

11. For literature see M. v. Ardenne, K. Steinfelder, and R. Tümmler, *Elektronenanlagerungs-Massenspektrographie organischer Substanzen*, Springer-Verlag, Berlin, 1971.

12. E. Gehrke and O. Reichenheim, *Verh. Dtsch. Phys. Ges., 8*, 559 (1906).

13. A. J. Dempster, *Nature, 135*, 542 (1935).

14. A. Benninghoven, *Surf. Sci., 35*, 427 (1973).

15. R. D. Macfarlane and D. F. Torgerson, *Int. J. Mass Spectrom. Ion Phys., 21*, 81 (1976).

16. Summarized in K. Biemann, *Mass Spectrometry*, McGraw-Hill, New York, 1962, p. 339.

17. G. Spiteller, A. Chatterjee, A. Bhottacharya, and A. Deb, *Monatsh. Chem., 93*, 1220 (1962).

18. P. Brown and C. Djerassi, *Angew. Chem., 79*, 481 (1967).

19. H. Budzikiewicz and M. Linscheid, *Org. Mass Spectrom., 9*, 88 (1974), and literature cited there.

20. T. W. Bentley and R. A. W. Johnstone, *Adv. Phys. Org. Chem., 8*, 191 (1970).

21. D. H. Williams and I. Howe, *Principles of Organic Mass Spectrometry*, McGraw-Hill, London, 1972, Chap. 6.

22. G. Remberg and G. Spiteller, *Chem. Ber., 103*, 3640 (1970).

23. H. Bosshardt and M. Hesse, *Angew. Chem., 86*, 256 (1974).

24. D. P. Stevenson, *Discuss. Faraday Soc., 10*, 35 (1951).

25. H. E. Audier, *Org. Mass Spectrom., 2*, 283 (1969).

26. H. M. Rosenstock, M. B. Wallenstein, A. L. Wahrhaftig, and H. Eyring, *Proc. Natl. Acad. Sci. USA, 38*, 667 (1952).

27. D. H. Williams and I. Howe, in Ref. 21, Chap. 4.

28. M. M. Bursey and F. W. McLafferty, *J. Am. Chem. Soc., 88*, 529 (1966).

29. J. L. Deroque, *Org. Mass Spectrom., 10*, 81 (1975).

30. D. H. Williams and I. Howe, in Ref. 21, Chap. 2.

31. C. Krier, J. C. Lorquet, and A. Berlingin, *Org. Mass Spectrom., 8*, 387 (1974).

32. K. Hirota and Y. Niwa, *J. Phys. Chem., 72*, 5 (1968).

33. R. C. Dougherty, *J. Am. Chem. Soc., 90*, 5788 (1968).

34. F. P. Boer, T. W. Shannon, and F. W. McLafferty, *J. Am. Chem. Soc., 90*, 7239 (1968).

35. J. H. Beynon, G. R. Lester, and A. E. Williams, *J. Phys. Chem.*, *63*, 1861 (1959).

36. K. Biemann, J. Seibl, and F. Gapp, *J. Am. Chem. Soc.*, *83*, 3795 (1961).

37. H. Budzikiewicz, C. Djerassi, and D. H. Williams, *Interpretation of Mass Spectra of Organic Compounds*, Holden-Day, San Francisco, 1964.

38. E. Murad and M. G. Inghram, *J. Chem. Phys.*, *40*, 3263 (1964).

39. H. Budzikiewicz and R. Pesch, *Org. Mass Spectrom.*, *9*, 861 (1974).

40. R. Stolze and H. Budzikiewicz, *Org. Mass Spectrom.*, *13*, 25 (1978).

41. J. H. Beynon, in Ref. 1a, P. 271.

42. H. Budzikiewicz, C. Djerassi, and D. H. Williams, *Mass Spectrometry of Organic Compounds*, Holden-Day, San Francisco, 1967, p. 159.

43. K. Biemann and G. Spiteller, *Tetrahedron. Lett.*, 299 (1961); *J. Am. Chem. Soc.*, *84*, 4578 (1962).

44. H. Budzikiewicz, C. Djerassi, and D. H. Williams, in Ref. 42, p. 490.

45. H. M. Grubb and S. Meyerson, in Ref. 7, Chap. 10.

46. R. G. Crooks, J. H. Beynon, R. M. Caprioli, and G. R. Lester, *Metastable Ions*, Elsevier, Amsterdam, 1973.

47. J. A. Hipple, R. E. Fox, and E. U. Condon, *Phys. Rev.*, *69*, 347 (1946).

48. H. Budzikiewicz, F. v. d. Haar, and H. H. Inhoffen, *Liebigs Ann. Chem.*, *701*, 23 (1967).

49. T. W. Shannon and F. W. McLafferty, *J. Am. Chem. Soc.*, *88*, 5021 (1966).

50. F. W. McLafferty and W. T. Pike, *J. Am. Chem. Soc.*, *89*, 5953 (1967).

51. J. H. Beynon and A. E. Fontaine, *Chem. Commun.*, 717 (1966).

52. J. Dickman, J. K. MacLeod, C. Djerassi, and J. D. Baldeschwieler, *J. Am. Chem. Soc.*, *91*, 2069 (1969).

53. D. F. Hunt, *Adv. Mass Spectrom.*, *6*, 517 (1974).

Chapter 2

APPLICATIONS OF MASS SPECTROMETRY IN PROTEIN STRUCTURE DETERMINATIONS

*ROBERT E. LOVINS**

Department of Basic and Clinical Immunology and Microbiology
Medical University of South Carolina
Charleston, South Carolina

*Current affiliation: Heart Research Institute, Institutes of Medical Sciences, San Francisco, California

I. INTRODUCTION

The sequences of amino acids in proteins contain a wealth of chemical, biological, and genetic information. Chemically, the information specifies the complex folding of the polypeptide chain, thus determining the biological function of the protein. Genetically, comparative studies of a single protein may make it possible to construct detailed phylogenetic trees for use in further studies of molecular genetics. Additionally, aberrations in protein structure responsible for many genetically inherited diseases are often identifiable from the elucidation of protein structure, and methods of treatment can be developed from this information. In view of the wealth of information contained in the amino acid sequences of proteins, it is not surprising that tremendous advances in methods of protein structure determination have been made during the past 15 to 20 years. Progress has been so rapid that the determination of the primary structure (amino acid sequence) of proteins has become one of the major areas of biochemical research.

In recent years mass spectrometry has been applied with increasing frequency as one of the tools used in probing protein structure. Biemann and co-workers [1-3] were among the first to recognize that amino acid sequence information could be extracted from the fragmentation patterns contained in the mass spectra of N- and C-terminally blocked small peptides. Since then, numerous investigations have established the mass spectral properties of a multitude of synthetic peptides as their N- and C-terminal blocked derivatives [4-8]. The elucidation of the structure of the natural peptidolipid fortuitin [9], using mass spectral techniques, was one of the first applications of mass spectrometry to the structure determination of a peptide of unknown structure. Improvements in the analysis of peptides by mass spectrometry have been made in the form of N-methylation of the amide nitrogens of the peptide bonds, thereby increasing the volatility of the peptide derivatives and extending the application of mass spectrometry to the analysis of peptides which would otherwise be too nonvolatile for mass spectral analysis [10,11].

Mass spectrometry has also been applied to the investigation of protein structure in concert with existing chemical techniques, in which case mass spectrometry has served primarily as an analytical tool to analyze amino acid derivatives obtained by the chemical degradation of proteins. In this regard mass spectrometry has been used to identify the N-terminal amino acids of proteins as their DNP [12,13] and DNS [14,15] derivatives.

Because of the almost universal acceptance of the Edman sequential degradation procedure [16] as the method of choice in biochemistry for the elucidation of protein and peptide structure and the recent automation of the technique, the application of mass spectrometry to the analysis of the products (phenylthiohydantoin derivatives) of the Edman reaction constitutes another major area of application of mass spectrometry in protein structure determination. The mass spectral behavior of all of the amino acid phenylthiohydantoin derivatives has been studied in detail, and criteria for their identification by mass spectrometry have been established [17-25]. Mass spectrometry is now being used with increasing frequency in concert with automated Edman sequencing techniques for the elucidation of protein structure [21,22,24,26].

In the following sections, the applications of mass spectrometry to the various areas of protein structure determination are outlined with emphasis on equipment, procedures, and protocols which have been established. By necessity, a large volume of work related to the mass spectral investigation of synthetic peptides of known structure and a number of studies involving theoretical aspects of mass spectrometry as applied to protein chemistry have been omitted.

II. MASS SPECTROMETRY IN PROTEIN END-GROUP ANALYSIS

The determination of the number of peptide chains constituting the protein (i.e., number of subunits) and the homogeneity of the protein preparation are among the first steps necessary in structural investigation. This information is normally obtained by determining the number of N-terminal amino acids present in the protein or peptide preparation. The number of different N-terminal residues combined with an estimate of the molecular weight provides an indication of the number of polypeptide chains present. Two methods of obtaining the N-terminal information will be discussed in this section. The first method involves the dinitrophenylation and subsequent isolation of the N-terminal amino acid of the protein or peptide as described by Sanger [27], and the second involves the dansylation and consequent isolation and identification of the DNS amino acid derivatives [28].

A. End-Group Analysis by Mass Spectral Identification of Dinitrophenylated Amino Acids

The N-terminus and exposed-side-chain primary amino groups of peptides and proteins are dinitrophenylated by treatment with 1-fluoro-2,4-dinitrobenzene using basic conditions that have been well documented [27]. The resulting DNP-peptide or protein is hydrolyzed subsequently in dilute acid [29], and the DNP-amino acid (N-terminal amino acid) is extracted into an organic solvent (usually chloroform or diethyl ether) and identified. Mass spectrometry offers several advantages over the other methods of identifying the DNP-amino acids (i.e., paper [30] and thin-layer chromatography [31]), such as unequivocal identification, sensitivity, and speed of analysis. The mass spectra of most of the DNP-amino acids [13] as well as their methyl esters [12] have been reported. Penders and co-workers [12] investigated the mass spectral behavior of the methyl esters of 17 DNP-amino acid derivatives. With the exception of the DNP derivatives of arginine, lysine, histidine, and tyrosine, all of the amino acids studied produce relatively intense molecular ions (M^+) and amine fragments (M^+-$COOCH_3$ ions). In the cases where no intense M^+ or amine fragment (M^+-$COOCH_3$) is present, other characteristic peaks are present which permit the amino acids to be distinguished in the presence of other DNP derivatives.

More recently Studier and Moore [13] investigated the use of mass spectrometry in conjunction with paper chromatography to identify DNP-amino acid derivatives. This application was developed for the analysis of amino acid mixtures generated in abiotic synthesis [13]. Table 1 shows the molecular ions and their relative abundance and the two most abundant fragment ions found in the mass spectra of several DNP-amino acids. In many cases the prominent peak is due to the amine fragment, DNP-NH-CH-R (M^+-COOH). Exceptions to this include DNP-phenylalanine,

TABLE 1

Mass Spectra of DNP-Amino Acids[a]

DNP-amino acid	M^+ m/e	Fragment ions m/e	Fragment ions m/e
Glycine	241 (21)[b]	196 (100)[b]	104 (18)[b]
Alanine	255 (19)	210 (100)	164 (26)
Serine	271 (41)	226 (100)	241 (48)
Proline	281 (13)	236 (100)	190 (17)
Valine	283 (16)	238 (100)	134 (23)
Leucine	297 (13)	252 (100)	196 (28)
Isoleucine	297 (13)	252 (100)	69 (23)
Phenylalanine	331 (10)	91 (100)	240 (24)
Glutonic acid	313 (7)	250 (100)	222 (83)
Tryptophan	370 (1)	130 (100)	43 (24)
Tyrosine	513 (0)	167 (100)	273 (56)
Histidine	487 (0)	247 (100)	276 (25)

[a]Data taken from M. H. Studier et al., *Biochem. Biophys. Res. Commun., 40*, 896 (1970).
[b]Numbers in parentheses are relative abundance values.

which is characterized by a base peak at m/e = 91 (tropylium ion), DNP-β-alanine, whose spectrum contains the base peak at m/e = 196 corresponding to the loss of CH_2COOH, and DNP-tryptophan, whose spectrum exhibits the most intense ion at m/e = 130. The spectrum of di-DNP-tyrosine has a base peak at m/e = 167, probably due to the DNP group itself, in addition to characteristic ions at m/e 273 and 240 resulting from fragments generated by splitting of the molecule between the α and β carbon atoms.

DNP-arginine and DNP-lysine suffer extensive decomposition during mass spectral volatilization, and their mass spectra are characterized by low-mass ions at m/e = 86 and 186, respectively. The other basic amino acid, di-DNP-histidine, produces a mass spectrum containing two characteristic ions, at m/e = 247, due to the DNP-imidazolyl methyl ion and at m/e = 276 due to loss of CO_2 and a DNP group. As in the case of the methyl esters, each DNP derivative exhibits mass spectral behavior which is unique and which can be used to identify unequivocally the derivative produced from an unknown N-terminal amino acid mixture.

The DNP derivatives can in most cases be unambigously identified, with the exception of DNP-leucine and DNP-isoleucine, by mass spectrometry and the analysis is considerably more rapid, less laborious, and more sensitive than the more common alternatives of thin-layer or paper chromatography. This method has been used in the author's laboratory to identify the N-terminal amino acids of the proteolytic enzyme trypsin isolated from shrimp, bovine, and human sources. In each case the enzyme was dinitrophenylated and hydrolyzed in 6 N hydrochloric acid, and the N-terminal DNP amino acid was extracted from the hydrolysate in chloroform and identified by mass spectrometry using the criteria outlined above (unpublished result).

B. End Group Analysis by Mass Spectral Identification of Dansylated Acids

The fluorescent reagent 1-dimethylaminonapthalene-5-sulfonyl chloride (dansyl chloride, DNS-Cl) is now widely used for the determination of N-terminal amino acids of proteins and peptides [28]. In using this procedure, free primary amino groups are coupled with the dansyl chloride reagent in alkaline solution. The DNS-amino acid derivatives thus formed are resistant to acid hydrolysis, allowing them to be isolated intact from the complete acid hydrolysis of the treated protein or peptide. The dansyl method plays an important role in sequence determination and to a considerable extent is replacing the DNP method described above. The particular advantage is the great sensitivity of the method. The DNS-amino acid derivatives have an intense yellow fluorescence, which allows their detection in micro quantities (0.1-0.01 nmol). The identification of DNS-amino acid derivatives has included primarily electrophoretic [32] and thin-layer-chromatographic [33] techniques, which in some cases do not provide completely unambiguous identification.

As in the case of DNP-amino acid derivatives, mass spectrometry offers a rapid, unambiguous alternative method for identifying DNS-amino acid derivatives. Marino and Buonocore [15] reported the mass spectral behavior of the methyl esters of 15 DNS-amino acids. With the exception of DNS-arginine and bis-DNS-cysteine, all of the methyl esters gave excellent spectra containing molecular ions with relative intensities ranging from 30% to 70% of the base peak, which in most cases was m/e = 170 (dimethylaminonapthyl ion). They also reported that the mass spectra obtained at a lower ionizing electron energy (12 eV) produced spectra containing in most cases only the molecular ion.

TABLE 2

Mass Spectra Data for DNS-Amino Acids[a]

DNS-amino acid	M^+ m/e	Fragment ions m/e	m/e
Glycine	308 (25)[b]	235 (4)[b]	250 (3)[b]
Alanine	322 (94)	235 (7)	287 (3)
Serine	338 (12)	235 (7)	250 (3)
Proline	348 (23)	235 (4)	303 (2)
Valine	350 (15)	235 (12)	250 (7)
Threonine	352 (9)	235 (11)	250 (7)
Leucine	364 (200)	235 (9)	250 (4)
Isoleucine	364 (87)	235 (13)	250 (12)
Asparagine	265 (2)	348 (15)	261 (15)
Aspartic acid	366 (74)	348 (90)	250 (6)
Lysine (ε-DNS)	379 (40)	235 (24)	250 (15)
Glutamine	379 (40)	235 (12)	297 (3)
Glutamic acid	380 (1)	235 (17)	362 (15)
Methionine	382 (22)	235 (12)	322 (6)
Phenylalanine	398 (200)	250 (41)	261 (10)
Arginine	407 (100)	267 (2)	265 (3)
Tryptophan	437 (220)	235 (270)	130 (200)
Tyrosine (O,N-bis-DNS)	647 (31)	235 (46)	414 (70)

[a]Data taken from Ref. 14.
[b]Numbers in parentheses are relative abundance values.

More recently Seiler and co-workers [14] have reported the mass spectral behavior of the free acid form of DNS-amino acids. Table 2 contains a compilation of the molecular ions (M^+) of the DNS derivatives with their corresponding relative abundance and two of the most intense fragment ions. As in the case of the methyl esters, the most intense ion in these spectra is at m/e 171, which is due to dimethylaminonapthalene. It is apparent, as in the case of the methyl esters, that the spectra of most of the DNS derivatives studied contain readily observable molecular ions, which in many cases constitute the base peak of the spectrum. The authors observed that the intensity of the molecular ions is a function of the temperature of the inlet system, as well as its structure. Low-intensity molecular ions are

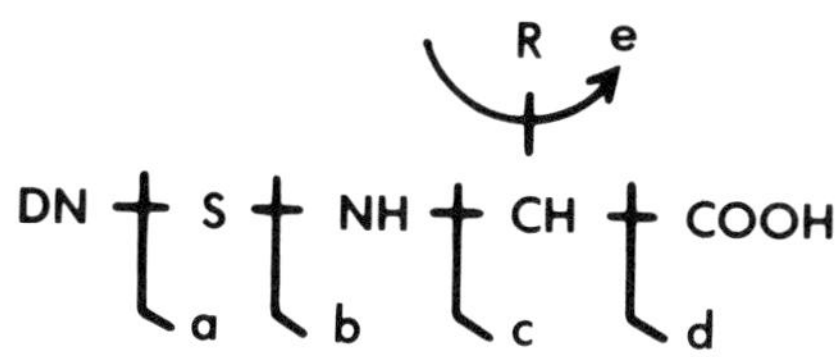

FIG. 1. Points of bond cission giving rise to major ions in the mass spectra of DNS-amino acid.

observed for the DNS derivatives of glutamic acid, glutamine, and asparagine. These derivatives tend to thermally eliminate H_2O or NH_3 during mass spectral analysis.

In general the fragmentation of the DNS-amino acid derivatives and their methyl esters are quite similar. In addition to the formation of dimethylaminonapthalene (m/e 171) (cleavage a in Fig. 1) and DNS^+ (m/e 235) (cleavage b), the major fragment is due to loss of the amino acid side chain R (cleavage e) (see Fig. 1).

β-Cleavage of the amine fragment (cleavage c) does not occur to any major extent, and similarly the loss of COOH (cleavage d) also occurs only to a slight degree.

Because of the lack of intense fragment ions, the DNS derivatives of leucine, isoleucine, norleucine, and hydroxyproline cannot be distinguished by their mass spectra. Consequently, in some cases, prior isolation by paper chromatography may be necessary for unambiguous identification. The best alternative to the use of mass spectrometry for identifying DNS-amino acids may be the equally sensitive fluorescent chromatographic techniques.

III. APPLICATIONS OF MASS SPECTROMETRY IN THE DETERMINATION OF TERMINAL AMINO ACID SEQUENCES OF PROTEINS AND PEPTIDES

After the number of subunits and identity of the N-terminal amino acids have been established, one of the next steps in elucidating the primary structure of a protein or peptide involves the determination of the amino acid sequences of the N- and C-terminal regions of the polypeptide chains. This information permits the investigator to recognize the N- and C-terminal peptides arising from the subsequent proteolytic and/or chemical degradation of the polypeptide chain into peptides to be used for further sequence analysis. Depending on the amino acid composition of the N-terminal portion of the peptide chain, two methods have been developed for the determination of the N-terminal amino acid sequence of the polypeptide chain.

A. N-Terminal Sequence Analysis of Proteins and Peptides Using the Edman Stepwise Degradation and Mass Spectrometry

As noted earlier, one of the most significant advances made toward solving problems of protein structure was Edman's development of the sequential degradation of a polypeptide chain from the N-terminus [16]. This procedure, referred to as the "Edman reaction" has proven to be an extremely useful technique especially with the advent of the commercially available automatic protein sequencer using the original Edman protocol [34].

The first step in the Edman Scheme is the formation of the phenylthiocarbamyl peptide from the coupling of the peptide with phenylisothiocyanate at pH 9 as shown in step I of the Edman reaction outlined in Figure 2.

The 2-anilino-5-thiazolinones formed by cleavage of the terminal phenylthiocarbamyl peptide by anhydrous acid, usually trifluoroacetic acid or heptafluorobutyric acid, are too unstable for identification by conventional chromatographic techniques and are usually converted to the more stable isomeric 3-phenyl-2-thiohydantoin (PTH). Except for serine and threonine PTH (Fig. 3a), which dehydrate upon heating to form the corresponding dehydro forms (Fig. 3b), resulting in subsequent polymerization of the unsaturated derivatives, all the PTH amino acids are relatively stable.

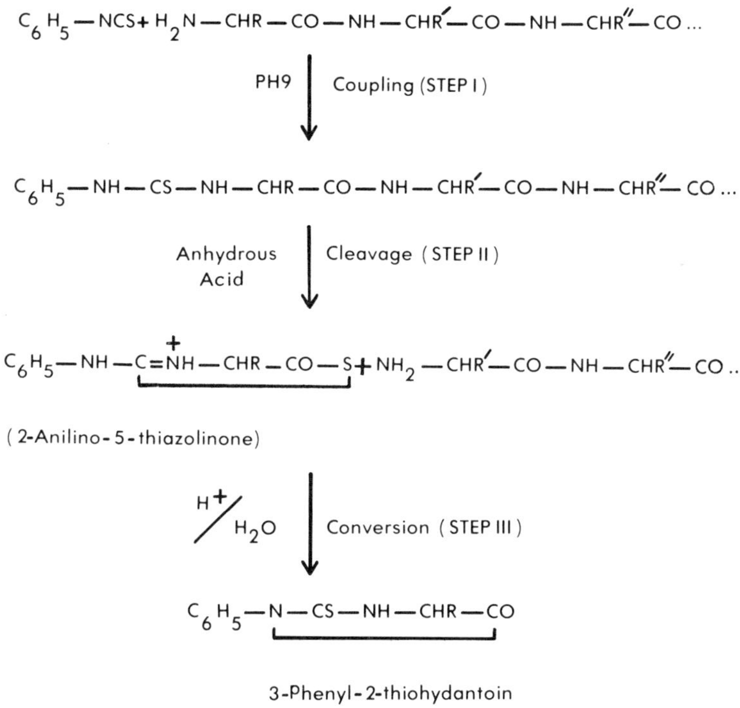

FIG. 2. Flow chart outlining the steps associated with the Edman degradation of a protein or peptide N-terminus.

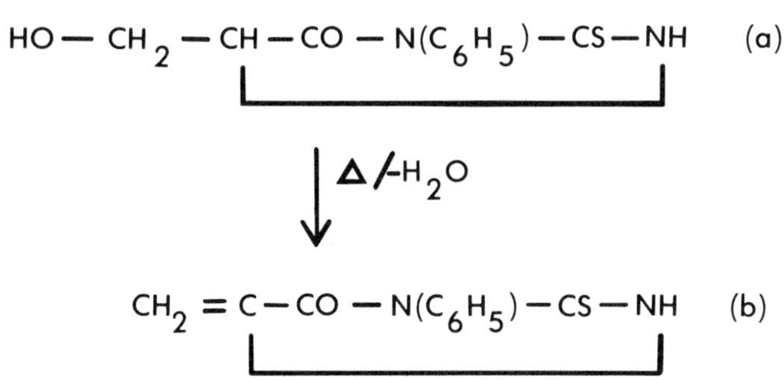

FIG. 3. The dehydration of serine PTH.

Not only is the thiazolinone intermediate unstable, the precursor thiocarbamyl derivative is susceptible to oxidative desulfuration to form the stable phenyl carbamyl peptide, which does not undergo further degradation. Consequently, precaution must be taken to exclude oxygen and other oxidizing agents from the reaction mixture in order to assure a high repetitive yield of amino acid derivative at each step of the degradation.

Gas-liquid chromatography (GLC) has gained wide acceptance as an analytical method providing relatively unambiguous analysis of the PTH-amino acid derivatives [35]. Because of the low volatility of many of the PTH-amino acids, however, volatile derivatives must be formed in order to facilitate their analysis by GLC methods. The most commonly used method of modifying the PTH derivatives for GLC analysis is the formation of their trimethylsilyl (TMS) derivatives. Except for arginine PTH, all of the amino acid PTH derivatives are amenable to GLC analysis as their TMS derivatives; however, serine and threonine are isolated as multiple peaks on GLC in reduced yields, and many times are indistinguishable from artifacts of silylation and from the reaction mixture background.

Other methods which have been used extensively for the analysis of PTH-amino acids include thin-layer chromatography on polyamide plates [36] and back hydrolysis of the PTH derivative to the free amino acid using HI or alkali with the subsequent identification of the free amino acid by amino acid analysis [37]. Not only are these methods time consuming but the back hydrolysis in many cases results in reduced yields of amino acids, especially acid-sensitive amino acids such as serine, threonine, and tryptophan.

With the development of the automatic Edman sequencer, resulting in increased speed, efficiency, and yield of products from the reaction, the need for methods to rapidly and unambiguously identify the products of the Edman reaction become more apparent.

TABLE 3

Mass Spectra Data for Amino Acid Phenylthiohydantons[a]

PTH-amino acid	Molecular ion	Relative intensity of M^+	m/e (base peak)	m/e	% Rel. abundance	m/e	% Rel. abundance	m/e	% Rel. abundance	Temperature [°C]
PTH-Gly	192	100	192	77	65	135	62	120	28	75
PTH-Ala	206	100	206	87	26	147	20	120	16	100
PTH-Val	234	100	234	192	55	205	17	219	4	90
PTH-Leu	248	98	135	192	52	205	25	219	18	90
PTH-Ile	248	97	192	219	20	57	14	--	--	110
PTH-Ser	222	30	77	204	26	119	14	84	12	125
PTH-Thr	236	12	192	218	28	163	8	120	15	130
PTH-Asp	250	100	250	204	28	234	12	177	9	125
PTH-Asn	249	100	249	205	54	85	23	44	23	110
PTH-Glu	264	100	264	248	40	204	40	119	21	135
PTH-Gln	263	21	77	59	69	205	33	219	6	150
PTH-CysSO$_3$H	286	--	64	64	100	48	56	204	30	150
PTH-CysCM	296	1	47	47	100	92	26	278	3	150
PTH-Met	266	100	266	205	83	61	41	--	--	110
PTH-Arg	291	1	93	155	40	127	24	249	4	130
PTH-Lys	398	--	135	93	38	263	18	246	6	115
PTH-Phe	282	58	91	136	21	91	100	249	5	115
PTH-Tyr	298	11	107	107	100	192	53	298	11	150
PTH-Trp	321	8	130	130	100	107	30	103	16	160
PTH-His	272	100	272	81	68	153	31	95	18	160
PTH-Pro	232	62	135	69	61	85	16	203	9	80

[a]Data taken from Ref. 25.

For this purpose, the mass spectral behavior of the amino acid PTH derivatives have been extensively documented, and criteria for the unequivocal identification by mass spectrometry have been established.

The mass spectral characterization of several aliphatic PTH-amino acid derivatives was first reported by Stepanov and Kristsov [38] and Melvas [39]. In 1970 Hagenmaier and co-workers [25] characterized the mass spectra of all the PTH-amino acid derivatives. With the exception of the PTH derivatives of serine, threonine, cysteic acid, arginine, and lysine, the mass spectra of all the derivatives exhibit relatively abundant molecular ions, which in many cases constitute the base peak of the spectrum. Table 3 contains the mass spectral data for the PTH derivatives. In those cases where no molecular ion is present, fragment ions are present which can be used for their identification. The major fragment ions observed in the spectra arise from the loss of the R group from the thiohydantoin ring, shown as fragment a in Figure 4, and also from the formation of C_6H_5NCS by fragmentation b (Fig. 4).

For derivatives in which R contains an aromatic group, as in phenylalanine and tyrosine, the base peak is due to the fragment ion corresponding to the R group (i.e., m/e 91, 107). The

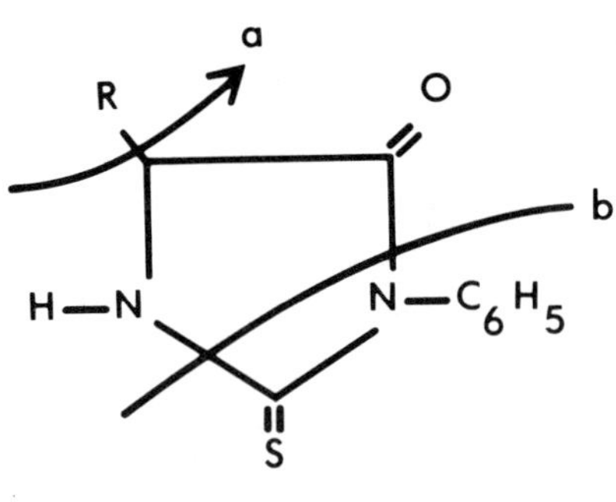

FIG. 4. The major fragmentation pathways found in the mass spectra of PTH-amino acid.

spectra of serine and threonine PTH contain ions due to M^+-H_2O (loss of water from the R group). More recently the mass spectral behavior of the PTH derivatives has been further studied using low ionizing voltages [23] and chemical ionization (CI) techniques [21]. With both techniques the molecular ion (MH^+ ion in the CI mode) is enhanced and the spectra are considerably simplified. Because of the advantages of speed and sensitivity afforded by this technique for analyzing the products of the Edman degradation reaction, three criteria for the identification of the phenylthiohydantoin derivatives derived from the Edman reaction from their mass spectra have been established [40]: (a) the mass of the molecular ion (M^+), (b) the relative intensity ratio of the molecular ion to one or two principal fragment ions, and (c) the probe temperature at which the PTH derivative volatilizes.

As shown in Table 3, the PTH derivatives can be divided into three groups. The first group includes the PTH derivatives of glycine, alanine, valine, leucine, isoleucine, and proline. These derivatives volatilize at probe temperatures below 105°C. With the exception of glycine, the mass spectra of each of the PTH derivatives in this group contain unique molecular ions which can be used to unambiguously identify their presence. Isoleucine and leucine can be differentiated by the presence of a fragment ion at m/e 205 in the spectrum of PTH leucine which is absent in that of PTH isoleucine. Further differentiation can be made on the basis of differing (M^+)/(m/e 192) ratios.

The second group of PTH derivatives, emerging in the temperature range 105 to 150°C, includes the PTH derivatives of serine, threonine, aspartic acid, methionine, glutamic acid, lysine, and phenylalanine. All of these derivatives produce spectra containing a recognizable and unique molecular ion, with the exception of lysine which has fragment ions at m/e 93, 135, 146, and 263, serine which has fragment ions at m/e 204 (M^+-H_2O) and 192, and threonine which exhibits a very weak molecular ion plus intense fragment ions at m/e 218 (M^+-H_2O) and 192. These fragment ions can be used for the mass spectal identification of the derivatives by measurement of their relative abundance as a function of the molecular ions of derivatives containing the same fragment ions.

The last group corresponds to the derivatives emerging above 150°C and includes the PTH derivatives of histidine, tryptophan, arginine, tyrosine, asparagine, and glutamine. The mass spectra of all the PTH derivatives contain unique molecular ions, with the exception of glycine, valine, asparagine, glutamine, and proline. The positive identification of each of these derivatives with molecular ions of the same mass as fragment ions appearing in the spectra of other thiohydantoin derivatives can be made on the basis of the temperature range at which they appear in the spectrum and/or the relative abundance of their molecular ion to a correspondingly abundant fragment ion which appears in their spectrum. For example proline, with a molecular ion at m/e 232, can be identified in the presence of asparagine whose spectrum contains a fragment ion at m/e 232. This can be done by measuring the relative intensity of m/e 232 to m/e 203. Both ions appear in the spectrum of proline and asparagine. The identification of the phenylthiohydantoin of valine in the presence of aspartic acid (the M^+ of valine at m/e 234 overlaps a fragment ion at m/e 234 in the spectrum of aspartic acid) can be made not only on the basis of the fact that valine volatilizes at a lower temperature range but also because the spectrum of valine contains two fragment ions at m/e 219 and 205, which are absent from the spectrum of aspartic acid.

In more recent investigations of the products of the Edman degradation reaction it was found that the 2-anilino-5-thiazolinone (phenylthiazolinone) derivatives, which are actually the species isolated from the Edman reaction, are amenable to mass spectral analysis, thus eliminating the third step of the procedure involving the acid-catalyzed conversion of the thiazolinone to the more stable thiohydantoin derivative [19]. The mass spectra of the phenylthiazolinone derivatives are essentially the same as those of the corresponding phenylthiohydantoin derivatives [19]. Apparently the thiazolinone derivatives undergo a thermal rearrangement to the more stable thiohydantoin derivative in the mass spectrometer, thus producing spectra characteristic of the thiohydantoin species. The ability to identify the thiazolinone derivatives directly without conversion to the thiohydantoin derivative offers several advantages. The acid-catalyzed conversion step is responsible for the partial destruction of acid-sensitive amino acids such as tryptophan and the loss of water from serine and threonine and their consequent polymerization. However, a problem of contaminating artifacts in the thiazolinone derivatives derived from the automated sequencer and their consequent confusion of the mass spectral identification of these derivatives has been found when the regular Quadrol buffer (N,N,N',N'-tetrakis-[2-hydroxypropyl]-ethylenediamine) is used in the sequencer program. Bauer and co-workers reported the use of 0.1 M Quadrol buffer for the automatic sequence analysis of small amounts of proteins and peptides [41]. The use of the dilute Quadrol maintains the same efficiency and high repetitive yields of thiazolinone as are obtained using the original 1 M Quadrol program. Using the dilute buffer system, the thiazolinones can be analyzed directly by mass spectrometry and spectra can be obtained which contain little or no background from sequencer artifacts.

Mass spectrometry has been used in concert with the automatic protein sequencer for the N-terminal sequence analysis of a number of proteins. In addition to the N-terminal sequence analysis of yeast enolase [42], the combined methods have been used to analyze the N-terminal sequence of two Bence-Jones proteins

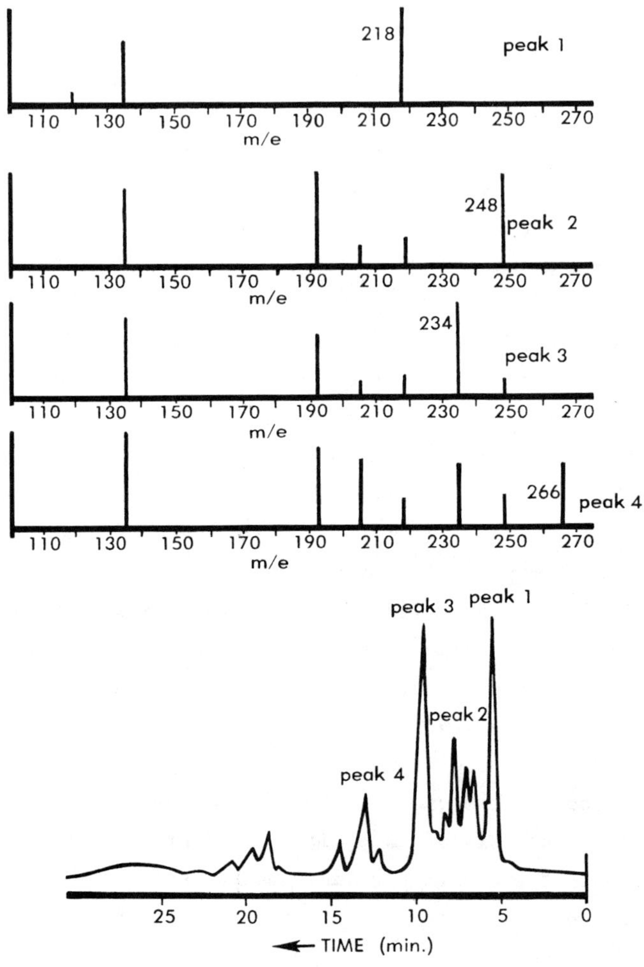

FIG. 5. The HPLC and corresponding mass spectra of the eluting HPLC peaks of the amino acid PTH derivative isolated at step 3 of the automatic N-terminus sequence analysis of anti-DNP antibody light chain BB-5.

derived from the urine of patients with multiple myeloma, as well as that of a hapten-specific antibody light chain and the polypeptide portion of the glycoprotein carcinoembryonic antigen (CEA) (unpublished results).

The use of mass spectrometry in concert with high-pressure liquid chromatography (HPLC) for the quantitative analysis of PTH derivatives from the automatic sequence analysis of proteins and peptides has also been reported recently [24]. In this application the PTH derivatives are separated on a silica gel matrix or by reverse phase chromatography using conditions developed by several investigators [43,44], and the eluting PTH peaks are collected and further analyzed by mass spectrometry using criteria described earlier in this section. A typical HPLC chromatogram and corresponding mass spectrum are shown in Figure 5. The advantages of this combined technique lie in the availability of two independent methods of identifying the phenylthiohydantoin derivative and the ability to quantitate the amount of each PTH from the liquid chromatogram. Using this technique, N-terminal sequence analyses of several antihapten antibody light chains of heterogeneous composition have been performed. The results of the quantitative N-terminal sequence analyses of two of these light chains are presented in Table 4.

B. N-Terminal Sequence Analysis of Blocked Proteins and Peptides Using Mass Spectrometry

While the majority of proteins to be investigated have free N-terminal amino acids, there are proteins whose N-terminus is composed of an amino acid whose amino group is blocked. Two major blocking groups found naturally are the acetyl group and pyrrolidone carboxylic acid. The N-terminus of the serum protein carbonic anhydrase, for example, is composed of N-acetylserine, while the N-terminus of the heavy chains of most immunoglobulins as well as many of the light chains of the same protein contain a pyrrolidone carboxylic acid group. The presence of an N-terminally blocked protein precludes the use of the stepwise Edman degradation method because of the necessity of a free N-terminal amino group to react with the Edman reagent, phenyl-

TABLE 4

Amino Acid Compositions of the N-Terminus of Early and Late DNP-ABG Light Chain Pool

Rabbit Number	Residue number[a,b] 1	2	3	4	5	6	7	8	9	10	11	12	13	14	15	16	17	18	19	20
BB-4	A(94)	V(50)	V(61)	M(74)	E(40)	Q(76)	T(ND)	P(68)	A(42)	S	V(100)	S	A(29)	A(54)	V(55)	S	G[c]	V(100)	V[c]	T(100)
P-1[d]	L(6)	Y(23)	T(25)	L(26)	M(36)	A(24)		A(32)	P(31)	A[c]		A[c]	V(28)	V(46)	A(45)	G[c]	V[c]		S	
		L(17)	M(14)		A(24)				G(27)	V[c]			L(28)			A[c]	L[c]			
		F(10)											T(15)				S			
BB-4	A(79)	V(58)	V(46)	M(48)	T(26)	Q(70)	T(64)	P(51)	A(82)	S	V(72)	S	A(45)	A(59)	L(29)	V(63)	V(100)	V(73)	V(100)	(ND)
P-4[d]	L(21)	L(18)	L(21)	T(32)	L(22)	G(30)	P(36)	A(49)	P(18)	A[c]	L(28)	A[c]	V(39)	T(41)	V(26)	G(37)		T(27)		
		Y(17)	T(20)	L(20)	M(17)								L(16)		G(22)					
		F(7)	M(7)		Z(17)										T(23)					
			G(6)		Y(9)															
					A(9)															

[a]The numbers in parentheses represent the percent of total PTH derivatives at each locus after subtraction of background contribution.
[b]The single letter notation for amino acids was used. A = ala, B = asx, C = cys, D = asp, E = glu, F = phe, G = gly, H = his, I = ile, K = lys, L = leu, M = met, N = asn, p = pro, Q = gln, R = arg, S = ser, T = thr, V = val, W = trp, Y = tyr, Z = glx
[c]Because the serine concentration is unknown, no attempt was made to quantitate these amino acids.
[d]P-1 or P-2 corresponds to pool of early antibody. P-3 or P-4 corresponds to pool 3 or 4 of late antibody.

isothiocyanate. Consequently, an alternative method for elucidating the N-terminal amino acid sequence of blocked proteins and peptides is necessary. Gray and del Valle described a method using mass spectrometry to elucidate the N-terminal amino acid sequence of N-terminally blocked proteins of peptides [45]. The first step in the procedure involves the exhaustive acetylation of the free α-amino groups of the lysine residues in the protein. The modified protein is digested with a protease of broad specificity, usually either the enzyme complex Pronase or Subtilisin, producing a mixture of peptides of which the N-terminal peptide is the only one not possessing an unblocked amino terminal. In order to increase the volatility of the peptides for mass spectral analysis, the peptide mixture is exhaustively N,O-permethylated using methyl iodide and methylsulfinyl carbanion (from reaction of dimethylsulfoxide and sodium hydride), in the manner described by Thomas and co-workers [10]. Other conditions for permethylating peptides and proteins include the use of silver oxide (Kuhn method) [46] and the method developed for O-permethylation of carbohydrates (proposed by Coggins and Benoiton [47]), which in the latter case is a faster technique, but results in many cases in C-methylation and other side reactions causing confusion in interpreting the mass spectral data of the peptide. To date the conditions described by Gray appear to be the method of choice for exhaustive N,O-permethylation of peptides and proteins. Leclercq and Desiderio [48] and Polan et al. [49] have, however, demonstrated that by using molar ratios of protein to carbanion to methyl iodide of 1:10:10, peptides containing cysteine, histidine, or methionine can be successfully permethylated without destruction of the sulfur-containing residues. All the peptides except the blocked peptide corresponding to the N-terminal portion of the protein are converted in the process to the quaternary ammonium salts, which remain in the aqueous phase of the final reaction mixture while the amino terminal peptide, now a neutral hydrophobic derivative, is partitioned into the organic phase (usually chloroform). The crude organic extract is subjected to mass spectral analysis using a mass spectrometer of mass range m/e 1000 and 1000 resolution (M/ΔM). The interpretation of the resulting mass spectra and reconstruction of the amino acid sequences from the spectra are based on the fact that the principal ions formed, apart from fragmentation of the amino acid side chains, arise from bond cission to form the sequence determining N-acyl peptide fragments. The permethylation of the peptides not only increases the volatility of the peptide, allowing the mass spectral analysis of larger peptides, but also results in simplified mass spectra, making the interpretation of the spectra more straightforward than those of acylpeptide esters whose amide nitrogens are not methylated. Because the N-methylation of the amide nitrogens in the peptide enhances bond cleavage at the amide linkage (as shown in Fig. 6), the N-acyl peptide

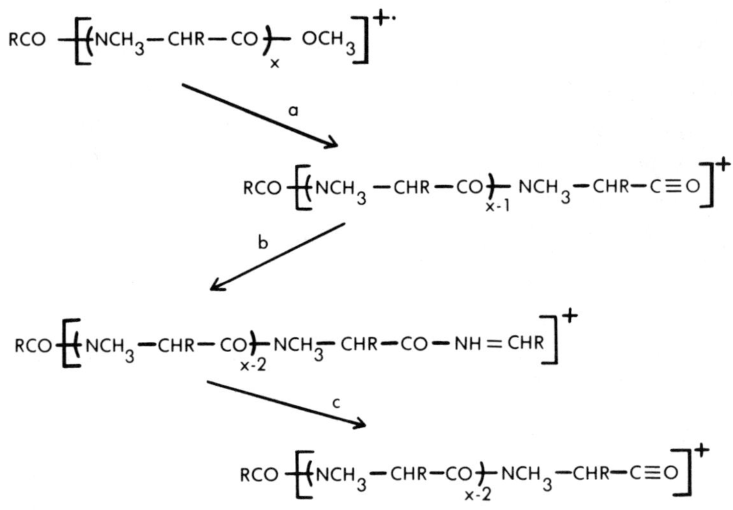

FIG. 6. The stepwise bond cission giving rise to the sequence determining ions found in the mass spectra of permethylated acetylated peptides.

TABLE 5

Major Sequence Ions in Spectrum of N-Terminal Peptide N-Acetyl Asp-Ile-Leu-Thr-OCH_4[a]

m/e	Amino acid residues
158	N-acetyl-asp (minus CO)
186	N-acetyl-asp
313	N-acetyl-asp-ile
426	N-acetyl-asp-ile-val
553	N-acetyl-asp-ile-val-leu
682	N-acetyl-asp-ile-val-leu-thr

[a]Data taken from Ref. 45.

fragments are in most cases the most abundant ions in the spectrum. Consequently, the N-terminal sequence can be deduced by identifying the first major ion (I_1) due to the acyl fragment (RCO^+) plus the N-terminal amino acid which would be 43 + 70, due to $N(CH_3)CH$-CO-, plus the mass of the side chain of the N-terminal amino acid (R). The next amino acid in the sequence is identified by identifying the second abundant peak in the spectrum whose mass is the sum of I_1, plus 70 amu due to the fragment $N(CH_3)CHCO$, plus the mass of the R group contained in the second amino acid. In this manner the N-terminal sequence of many terminally blocked proteins can be reconstructed. Table 5 shows the major sequence ions found for the N-terminal peptide N-acetyl Asp-Ile-Val-Leu-Thr-OCH_3.

This method has been used successfully to identify the N-terminal sequence of the enzyme carbonic anhydrase, acetyl-ser-pro-, from goat, human, porcine, and bovine sources (unpublished results). In addition, the method is being developed further to identify the N-terminal sequence of PCA-blocked immunoglobulin heavy chains. In those cases where the N-terminal peptide released from the proteolysis of the protein is larger than three or four amino acids, the relative intensity of the sequence-determining ions decreases with increasing mass, and the spectrum becomes indeterminant beyond masses corresponding to 10 or 11 amino acid residues. However, one is nevertheless able to follow the amino acid sequence from the N-terminal end into the peptide chain for several amino acids, providing N-terminal sequence information even though the spectrum of the complete peptide is not discernable. It should be noted at this point that even though the volatility of the peptide is greatly enhanced by exhaustive permethylation, the presence of the amino acid arginine prevents the volatilization of the peptide in the mass spectrometer, owing to the saltlike nature of the guanidino group. Consequently, arginine containing peptides must be modified prior to sequence analysis by one of the following techniques. If the N-terminal peptide, or any peptide for that matter, is the product of the trypsin digestion of a protein so that the arginine is the C-terminal amino acid, the

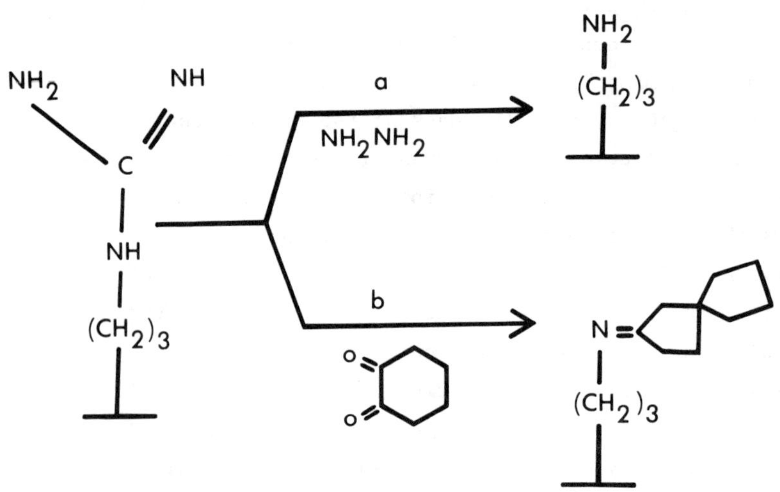

FIG. 7. The conversion of the guanidino group of arginine to (a) ornithine by reduction with hydrazine and (b) a spirocyclopentane derivative with cyclohexanedione.

peptide can be treated with carboxypeptidase to remove the C-terminal arginine prior to sequence analysis. For those arginine residues which are internal in the peptide, one can hydrolyze the arginine to ornithine by treatment with hydrazine [50] (Fig. 7a), or one can treat the arginine containing peptide with an α-diketone such as cyclohexanedione (Fig. 7b) to form the spirocyclopentane derivative. Hydrolysis with hydrazine results in some spurious cleavage of peptide bonds (10-60%) under the conditions necessary to convert arginine to ornithine and may cause interfering peptide fragments, resulting in ambiguous sequence assignment. Arginine residues in intact proteins can be converted to their spirocyclopentane derivatives using cyclohexanedione in greater than 90% yield with no apparent interfering side reactions. This procedure has been used to modify arginines in a large number of peptides, and their mass spectra have been well documented [51].

For the N-terminal sequence analysis of proteins having an acylated or otherwise blocked amino terminus, the procedure described by Gray and del Valle using mass spectrometry is proving to be a valuable alternative procedure and is providing N-terminal sequence information which would not otherwise be obtainable.

C. C-Terminal Amino Sequence Analysis of Proteins and Peptides Incorporating Mass Spectrometry

While the procedures for establishing the N-terminal sequence of a protein or peptide are well established and almost universally used in the determination of protein structure, the determination of carboxyl terminal (C-terminal) sequences has enjoyed only limited success. Considerable progress was achieved by Stark [52], who reinvestigated the formation of C-terminal peptidylthiohydantoins first proposed by Schlack and Kumpf [53]. By specific cleavage, the C-terminal amino acid can be removed as

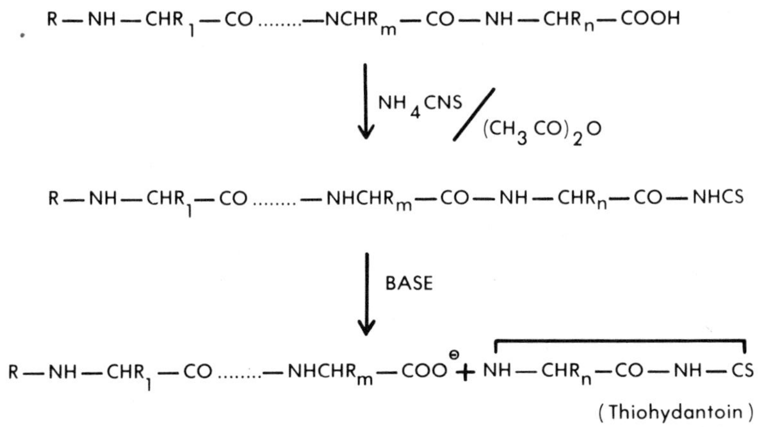

FIG. 8. Flow chart outlining the stepwise formation of thiohydantoins amino acid derivative from the C-terminus of a peptide or protein on treatment with ammonium thiocyanate.

its thiohydantoin derivative and identified. Peptides substituted at the N-terminal amino group are converted to the peptidyl thiohydantoin by heating for 30 min at 100°C with ammonium thiocyanate in acetic anhydride. The thiohydantoin corresponding to the C-terminal amino acid is released by treatment with base (0.1 M acetohydroxamate in 50% pyridine), according to the reaction scheme shown in Figure 8. The released thiohydantoin is extracted in organic solvent and identified. While not as efficient as the N-terminal Edman sequential degradation reaction, the C-terminal reaction does permit the sequential analysis of two to four amino acids with confidence.

As in the case of the substituted amino acid thiohydantoin derivatives derived from the N-terminal degradation procedures, the identification of the C-terminal thiohydantoin amino acids has relied largely on thin-layer or column chromatography [52]. Recently the use of gas-liquid chromatography and mass spectrometry to identify these C-terminal derivatives has been described [54]. The mass spectra of the thiohydantoins of 15 amino acids

TABLE 6

Mass Spectral Characteristics of Amino Acid Thiohydantoin Derivatives[a]

Amino acid	M^+ m/e	Fragment ions m/e	m/e
Glycine	116 (100)[b]	88 (20)[b]	60 (22)[b]
Alalanine	130 (100)	102 (20)	43 (40)
Valine	158 (70)	116 (100)	98 (15)
Isolevine	172 (70)	116 (100)	57 (20)
Leucine	172 (80)	116 (100)	129 (40)
Threonine	160 (0)	142 (100)	55 (60)
Methionine	190 (100)	129 (60)	116 (80)
Asparagine	173 (100)	128 (90)	156 (50)
Aspartic Acid	174 (100)	128 (80)	60 (60)
Glutonic Acid	188 (100)	128 (90)	170 (50)
Phenylalanine	206 (40)	91 (100)	--
Tyrosine	222 (5)	107 (100)	77 (15)
Tryptophan	245 (1)	130 (100)	77 (10)
Histidine	196 (75)	81 (100)	54 (12)
N-Azlys	229 (100)	187 (50)	170 (55)

[a]Data taken from Ref. 54.

have been reported. The mass spectral characteristics are outlined in Table 6. For GLC analysis, the thiohydantoin derivatives require prior conversion to their more volatile trimethylsilyl derivatives, a step not required for mass spectral analysis. Of the 15 derivatives studied, the mass spectra of all but three contained relatively intense molecular ions which can be used for their unambiguous identification. The spectrum of threonine thiohydantoin contained an intense ion at M^+-H_2O due to loss of water from the threonine R group. The major ions in the spectra of tyrosine and tryptophan thiohydantoin were m/e 107 and 130 due, respectively, to the aromatic tyrosine R group and the indole R group of tryptophan. The fragmentation of all of the thiohydantoin derivatives could be rationalized in terms of the loss of the substituent amino acid R group, followed by fragmentation of the thiohydantoin ring or by fragmentation of the ring with R attached. The stability of the thiohydantoin ring is demonstrated by the presence of an ion at m/e 116 in the spectra of many of the amino acid thiohydantoin derivatives. While no one has yet reported using mass spectrometry to identify the thiohydantoins released from the C-terminal stepwise degradation of a protein or peptide, the same potential for its application in this context exists as has been found for its use in identifying the PTH derivatives derived from the stepwise Edman degradation of the N-terminus of protein.

IV. STRATEGIES FOR PROTEIN STRUCTURE ELUCIDATION INVOLVING MASS SPECTROMETRY

Once the N- and C-terminal amino acid sequence of the protein or polypeptide has been established using one or more of the methods described in the previous section, the protein must be systematically fragmented into peptides of less complexity, using enzymatic and/or chemical means. In order to reconstruct the complete primary structure of a protein or large polypeptide from its peptide fragments, the methods used to cleave the protein chain must produce peptide fragments which represent overlapping regions of the original protein, so that the peptide sequence can be placed in the order of their appearance in the protein structure. To accomplish this, specific enzymes and chemical methods are usually chosen so that the cleavage of the polypeptide chain occurs at a restricted number of amino acid residues of known structure. For example, the proteolytic enzyme trypsin cleaves specifically at the C-terminal bond of arginine and lysine residues [55]. The specificity of trypsin can be decreased or broadened by chemically modifying the side chains of arginine and lysine as well as the amino acid cysteine prior to trypsinization of the protein. If the ε-amino groups of the lysine residues are chemically altered by treatment with maleic anhydride [56] or other acylating agents, peptides arising from cleavage solely at arginine residues result, while modification of the guanidino group of the arginine residues by treatment with cyclohexanedione or other α-diketone [51] restricts trypsin cleavage to lysine residues. By modifying both the lysine and arginine residues and forming the s-aminoethyl derivative of cysteine residues, trypsin will cleave only at the modified cysteine molecules in the protein [57]. Another example of specific cleavage involves the use of cyanogen bromide, which cleaves specifically at methionine residues to produce peptides containing a C-terminal homoserine lactone [58]. Other chemical procedures have been developed such as the N-bromosuccinimide cleavage of tryptophan residues [59] and the cleavage of acid-sensitive Asp-Pro bonds with acetic acid [60].

Partial acid hydrolysis of proteins and polypeptides in many cases results in the formation of peptides, many of which contain overlapping sequence information. The extent to which partial hydrolysis yields a mixture of peptides that contain sufficient overlapping of sequence information to be useful is largely dependent on the size of the proteins under investigation. Generally, the peptide mixtures obtained from a protein or larger polypeptide by partial acid hydrolysis are exceedingly complex and present special problems in separation and purification. However, the partial acid hydrolysis of a smaller peptide fragment in many instances provides a less complex mixture of peptides useful for reconstructing the sequence of the peptide.

The recent isolation of dipeptidylaminopeptidases (DAP), which catalyze the hydrolysis of polypeptides or proteins to dipeptides starting from the N-terminus of the polypeptide chain, provides an alternative to partial hydrolysis for generating a mixture of small peptides from a larger polypeptide chain [61]. By using a mixture of DAP I and IV, the restriction in specificity observed for DAP I alone, which does not hydrolyze peptides containing N-terminal arginyl or lysyl residues, is eliminated [62]. The only other restriction is that the enzyme cannot cleave the amide bonds of proline.

After the protein or polypeptide is fragmented by any of the above methods or by alternative methods, the resulting polypeptide mixture must be separated into its component parts prior to sequence analysis of the individual peptides. A number of separation procedures have been developed for this purpose. These include gel filtration [63], electrophoresis [64], and ion-exchange chromatography [65]. The techniques employed should result in a complete separation of the peptides to ensure that the analysis and sequencing of one peptide in the digest are not complicated by the presence of another.

It must be kept in mind that because proteins vary in their size, amino acid composition, and primary structure, the strategies developed for cleaving one protein and separating the resulting peptides may not be suitable for another protein. The choices of enzymes, for example, are determined by the number of arginyl, lysyl, cysteinyl, and hydrophobic residues present in the protein. Consequently, no single strategy will work universally.

After the protein has been fragmented into smaller peptides by the above methods, mass spectrometry can be utilized in a number of ways to help elucidate the structure of the peptide fragments. Nau has reported the use of combined gas-liquid chromatography and mass spectrometry (GLC-MS) for the separation and sequence analysis of mixtures of small di- and tripeptides such as would result from partial acid hydrolysis of a larger polypeptide chain [66]. The acetylated or perfluoroalkylated peptides are reduced with lithium aluminum deuteride to the perdeutero-N-acyl or N-perfluoroacylpolyamino alcohol. These derivatives and their methyl or ethyl esters are amenable to GLC analysis. Nau reports the enhancement of the perfluoroalkylated derivatives for GLC analysis by O-trimethylsilylation. In each case the modified peptides are separated by GLC and their sequences are identified from their mass spectra. The use of the more volatile O-TMS derivatives provides better GLC separation, permitting complete mass spectrometric characterization of more complex mixtures of peptides. While the procedure produces unambiguous results for peptide mixtures of known composition, its application to the sequence analysis of peptide mixtures from the partial acid hydrolysis of polypeptides of unknown structure has as yet to be demonstrated. The method has considerable potential when used in concert with other sequence strategies.

Another application of combined GLC-MS to peptide structure has been reported by Caprioli and Seifert [62], who analyzed dipeptide mixtures released upon hydrolysis of peptides and proteins with dipeptidylaminopeptidase I and IV. The separation of the dipeptides as their acyl ester derivatives, followed by sequence analysis by mass spectrometry, provides a rapid and convenient method for determining the structure of the dipeptide in the mixture. Because the peptidase is blocked by proline, the method is best suited for the degradation of polypeptides of known amino acid content rather than intact proteins. By using the DAP in concert with the Edman stepwise degradation, overlapping dipeptides can be generated from a larger peptide whose sequence analysis by this procedure permits at least partial reconstruction of the primary structure of the larger peptide. It has been demonstrated that for larger peptide chains proportionally greater numbers of overlapping peptides are necessary for unique definition of the structure of the peptide chain [67]. Using only overlapping dipeptides, the primary structure of the precursor peptide is more difficult to deduce because of the larger number of combinations possible for the overlapping dipeptides; consequently, in order to resolve the ambiguities it is necessary in many cases to generate larger overlapping peptides to provide the missing information. In some cases, however, a knowledge of the amino acid composition of the peptide is sufficient to allow the assignment of the correct sequence from the sequence analysis of the dipeptide mixture.

In many cases the sequences of the larger peptides released by proteolytic treatment of the protein, i.e., trypsinization, are determined without further hydrolysis or treatment. Using the automated protein sequencer, the sequence of these peptides can often be determined in its entirety. In these cases mass spectrometry is used to identify the PTH-amino acid derivatives released at each step of the stepwise automatic degradation of the peptide. Hydrophobic peptides are more difficult to sequence in the automatic sequencer because of peptide losses due to co-extraction of the peptide from the reaction vessel during the washing steps of the cycle. Braunitzer et al. have described a technique to partially eliminate this peptide loss by extraction [68]. The modification involves reacting the peptide with p-sulfonylphenylisothiocyanate. This reagent reacts with the N-terminal of the peptide and ε-amino groups of lysine residues in the peptide. The first step of the stepwise degradation subsequently releases the N-terminal amino acid as the sulfonic-acid-substituted thiazolinone, leaving the remaining peptide as a polar peptide due to the sulfonic acid groups attached to the ε-amino groups of the lysine residues. The procedure is obviously limited to lysine-containing peptides or peptides arising from trypsin cleavage of the protein.

For the smaller hydrophobic peptide (2 to 10 amino acids) the method described by Gray and del Valle [45] is applicable for sequence analysis. Acetylation of the peptide with an equimolar mixture of acetic anhydride and perdeuteroacetic anhydride followed by exhaustive N,O-permethylation as described in the previous section produces a peptide derivative which is directly amenable to mass spectral analysis. The mass spectra obtained for peptides modified in this manner contain N-acylpeptide fragment ions, i.e., sequence ions, which appear as "doublets" three mass units apart, arising from the use of the deuterated and non-deuterated acetic anhydride. The "doublet" sequence peaks make identification of the sequence-determining ions in the spectra much easier and less subject to interference from ions due to amino acid side-chain fragmentation. Figure 9 shows the spectrum of acetyl-asp-ile-val-leu-thrOCH_3 in which the doublet sequence-determining peaks are observable [45].

It may be necessary in the case of some peptides to use a combination of techniques in order to achieve the reconstruction of their primary structures. For example, many peptides derived from either proteolytic or chemical cleavage of a protein are directly amenable to the stepwise Edman degradation procedure using mass spectrometry to identify the PTH derivatives at each step of the cycle in an automatic protein sequencer for several cycles until the peptide chain is reduced in size to that which is amenable to direct mass spectral analysis, as their acetylated, permethylated derivatives. In this manner the best advantage is made of both techniques to obtain the amino acid sequence of the entire peptide.

The strategies discussed in this section are by no means exhaustive but are meant to serve as guidelines for identifying the many areas of protein sequence analysis to which mass spectrometry can and has been adapted. Many steps in the analysis

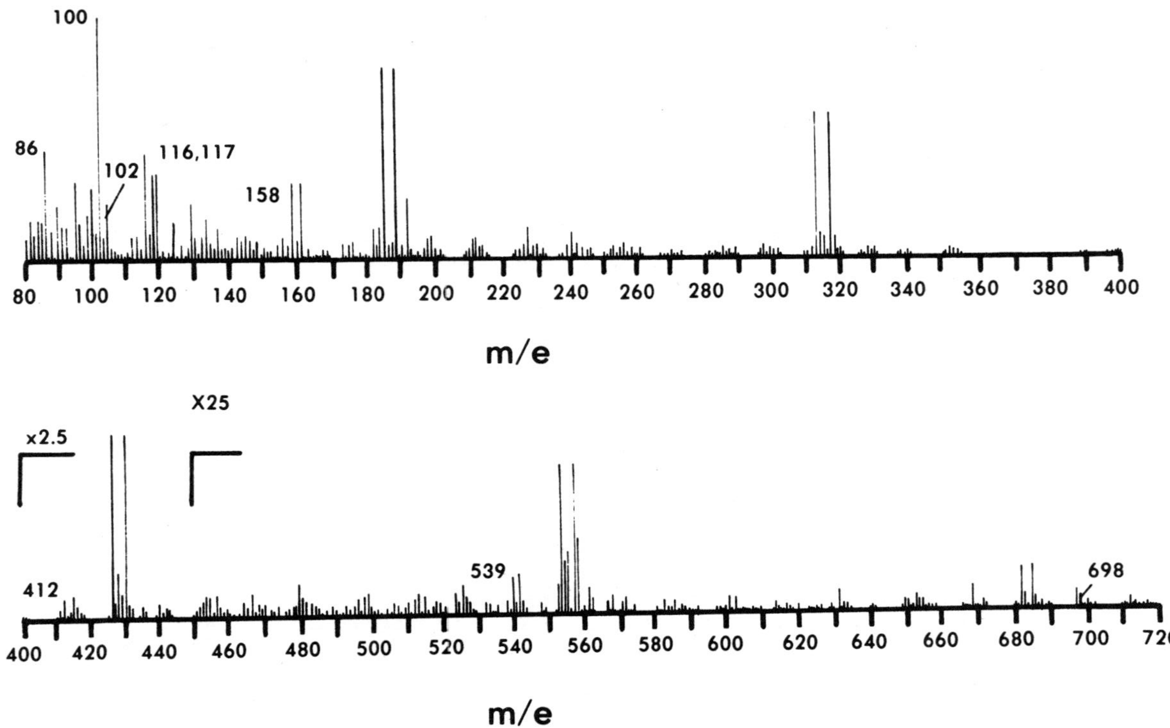

FIG. 9. Mass spectrum of a peptide permethylated and acetylated with a mixture of perdeutero acetic anhydride and unlabeled acetic anhydride giving rise to sequence determining ions separated by 4 amu. (Taken from Ref. 45.)

and characterization of the protein which do not directly involve mass spectrometry but which produce necessary information about the properties of the protein, i.e., amino acid composition, molecular weight, etc., have not been discussed. Similarly, many applications of mass spectrometry to the sequence analysis of specific peptides and proteins have not been included because the discussion was not meant to be a rigorous review of the uses of mass spectrometry in protein chemistry but rather to present the scope of application which can be made of mass spectrometry in the determination of the primary structure of a protein.

V. A STRATEGY FOR DETERMINING THE PRIMARY STRUCTURE OF A PROTEIN USING SPECIFIC PROTEOLYTIC AND CHEMICAL FRAGMENTATION AND QUANTITATIVE PROTEIN SEQUENCING BY MASS SPECTROMETRY

Inherent in all the strategies for determining protein structure are the isolation, separation, and complete amino acid sequencing of all the peptide fragments derived from the various protein fragmentation methods chosen. The separation, characterization, further cleavage, and/or protein cleavage to produce overlapping sequence information represent an arduous undertaking. It is with this in mind that this final section outlines a strategy for determining the sequence of a protein which is designed to circumvent some of the time-consuming areas of protein structure elucidation such as separation, characterization, and complete sequence analysis of each individual peptide in the mixture of peptide fragments derived from multiple cleavage of a protein.

The strategy to be described is based on that outlined by Gray [69], in which he proposed to determine the N-terminal sequence of mixtures of peptides, without prior fractionation, produced by a series of selective cleavages of the protein chain. Analysis of the amino compositions at each locus of the peptide mixture would provide a "nearest-neighbor" analysis extending successively to neighboring residues from the point of cleavage.

The success of the protocol depends on the quantitative identification of all the amino acids at each locus of the N-terminal sequence of the peptide mixture. The development of the automated protein/peptide sequencer [34], capable of stepwise degradation of a peptide or protein 20 or more amino acids from the N-terminal, and the successful application of mass spectrometry to the identification of the PTH-amino acid derivatives produced by the automatic stepwise degradation [25,26], make possible the quantitative identification of the N-terminal sequences of the peptide mixtures.

The quantitative identification of mixtures of PTH-amino acid derivatives by mass spectrometry using ^{15}N-isotope dilution techniques [40] involves the addition of known quantities of ^{15}N-labeled PTH-amino acids to the PTH mixture isolated at each step of the stepwise peptide degradation prior to mass spectral analysis. By accurate measurement of the $^{15}N/^{14}N$ ratios of the molecular and/or fragment ions of each PTH derivative in the mixture, the molar quantity of each derivative can be determined at each step in the degradation. High-pressure liquid chromatography (HPLC) and its use in conjunction with mass spectrometry provides an alternative to isotope dilution for quantitative identification of PTH derivatives [24]. Using silica gel columns or reverse phase columns, the PTH-amino acids can be separated and identified individually by mass spectrometry. The PTH-amino acids eluting from the HPLC column are detected by absorption at 254 nm, while mass spectrometry provides absolute, unequivocal identification of each derivative as it is eluted from the HPLC column.

The successful assignment of each amino acid at each locus of the peptide mixture depends not only on the quantitative identification of each PTH-amino acid but also requires that the protein be cleaved specifically so that the amino acids comprising the cleavage points are known. As an example, the fragmentation of the enzyme lysozyme, using proteolytic and chemical means to generate four sets of peptides derived by cleavage at known points, has been described [70,71]. Treatment of a portion of lysozyme with cyanogen bromide forms a mixture of peptides from cleavage at the two methionine residues. After modification of the guanidino group of the arginine residues of a second portion, the modified protein is treated with trypsin to produce a second mixture of peptides resulting from cleavage at the four lysine residues. Trypsin cleavage of a third portion of lysozyme modified by treatment with maleic anhydride or citroconic anhydride to block the ε-amino groups of the lysine residues produces a third mixture of peptides resulting from cleavage at each of the 11 arginine amino acids. The fourth peptide mixture is generated by blocking both the arginine and lysine residues and treating with trypsin after forming the S-aminoethyl derivatives of the eight cysteine residues.

The flow chart in Figure 10 outlines the modifications necessary to generate the four peptide mixtures. Each method of fragmentation must be chosen to produce peptides by cleavage at

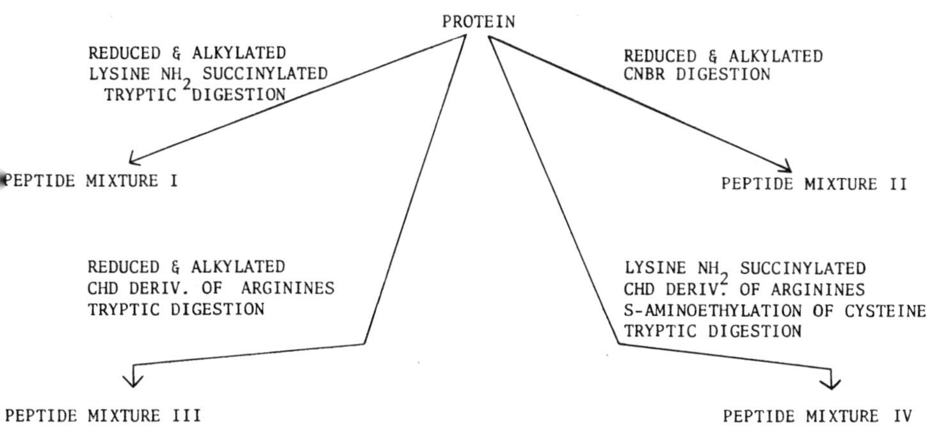

FIG. 10. Flow chart outlining the cleavages used to generate four discrete peptide mixtures of known origin.

TABLE 7

N-Terminal Amino Acid Composition of Peptide Mixtures Derived from Cleavage of Lysozyme as Indicated in Figure 10

Cleavage after	Steps of sequential degradation 1	2	3	4	5	6	7	8	9	10	11	12	13	14	15
(a) Met															
	Asn	Ala	His	Gly	Ala	Asp	Asn	Asn	Ala	Ala	Ala	Gly	Leu	Asp	Asn
	Lys	Arg	Phe	Gly	Arg	Cys	Arg	Leu	Arg	Cys	Lys	Ser	Thr	Gly	Val
	Lys	Val	Trp	Val	Leu	Trp	Glu	Tyr	Arg	Gly	Tyr	Met			
(b) Lys															
	Arg	Glu	Asp	Asn	Asp	Ala	Gly	Ala	Ala	Ala	Asn	Arg	Ala	Asn	Asp
	Ile	His	Gly	Asp	Cys	Asn	Leu	Arg	Ala	Ala	Cys	Arg	Asn	Thr	Arg
	Gly	Phe	Gly	Arg	Gln	Asn	Thr	Ile	Asn	Gly	Met	Leu	Gly	Trp	Trp
	Lys	Thr	Ser	Leu	Gly	Asp	Trp	Glu	Arg	Thr	Ser	Lys	Leu		
	Lys	Val	Ser	Val	Phe	Glu	Tyr	Met	Gly	Tyr	Trp	Val			
	Phe	Val													
	Val														
(c) Cys															
	Glu	Val	Ala	Ala	Arg	Met	Ala	Arg	Ile	Gly	Gly	Asp	Asn	Tyr	Arg
	Ala	Leu	Phe	Phe	Ala	Ser	Gly	Phe	Arg	Thr	Leu	Ala	Asn	Asn	Arg
	Ala	Ala	Lys	Gly	Glu	Ser	Asp	Gly	Thr	Gly	Ser	Asn	Ala	Trp	Val

Ser	Lys	Lys	Ile	Val	Ser	Asp	Ile	Asp	Ala	Met	Val	Thr	Cys
Asn	Ala	Leu	Leu	Ser	Pro	Asn	Ser	Asn	Asn	Gln	Cys		
Lys	Asp	Gly	Arg	Thr	Glu	Lys	Trp	His	Arg	Leu	Cys		
Asn	Gly	Thr	Asp	Val	Cys								
Arg	Ile	Pro	Cys										
Lys	Leu	Cys											

(d) Arg

Gly	Val	Arg	Gly	Gly	Asn	Trp	Val	Arg	Ala	Arg	Lys	Ser	Glu	Ile
Asn	Tyr	Gly	Leu	Arg	Thr	Asp	Tyr	Trp	Ile	Leu	Glu	Ile	Asn	Ser
Asn	Thr	Cys	Gly	Ser	Pro	Cys	Ser	Ala	Ile	Leu	Ser	Phe	Asp	Ser
Cys	Leu	Leu	Asn	Ile	Val	Glu	Ala	Gly	Leu	Ala				
Cys	Lys	Leu	Thr	Asp	Ala	Met	Lys	Cys						
His	Glu	Gly	Ala	Ala	Tyr	Arg								
Trp	Gly	Cys	Asp	Asn	Gly	Arg								
Thr	Trp	Asp	Asn	Asp										
Gly	Pro	Ser	Ser	Arg										
Asn	Cys	Phe												
Leu	Arg													
Lys	Arg													

only one kind of amino acid. Other methods of cleavage can be employed, but the ones outlined result in the most specific cleavages with the highest yields of peptides. The number and identity of the amino acids at each locus of the N-terminal 15 to 20 residues for each of the four sets of peptide mixtures are determined by stepwise degradation of the peptide mixtures and mass spectral identification of the PTH mixtures at each step of the degradation. The N-terminal sequence of each peptide in each mixture can be reconstructed by nearest-neighbor analysis of the data obtained for the four peptide mixtures.

Table 7 shows the N-terminal amino acid compositions found in the cases of the cyanogen bromide peptide mixture and the peptide mixture obtained from trypsin cleavage at lysine, as well as those which are expected from the quantitative N-terminal sequence analysis of the peptide mixture from cleavage at arginine and at cysteine. The first step in the reconstitution of the sequence data is to extract the amino acids at each locus of each peptide mixture, corresponding to the previously determined N-terminal sequence (Table 7): lys-val-phe-gly-arg-cys-glu-leu-ala-ala-ala-met-lys-arg-his. The second step involves comparing the remaining data in the four peptide mixtures, i.e., cys, met, lys, and arg. For example, comparing data sets (a) and (b) and knowing the N-terminal sequence, one can deduce the two remaining met sequences. This is accomplished by comparing column 2 in set (a) with column 1 of set (b). For example, only arg in 2a is found in 1b, only his in 3a is found in 2b, only gly in 4a is found in 3b, and leu in 5a is present in the mixture in 4b, and so on. Using these first-order comparisons, the N-terminal sequences of the two peptides derived from the cyanogen bromide cleavage of lysozyme are met-lys-arg-his-gly-leu-asp-asn-tyr-ser-leu-gly-asn- and met-asn-ala-trp-val-ala-trp-arg-asn-arg-cys-lys-gly-thr-asp-val-. The first of the two sequences accounts for one of the sequences in the lysine set (b) as a result of cleavage at a

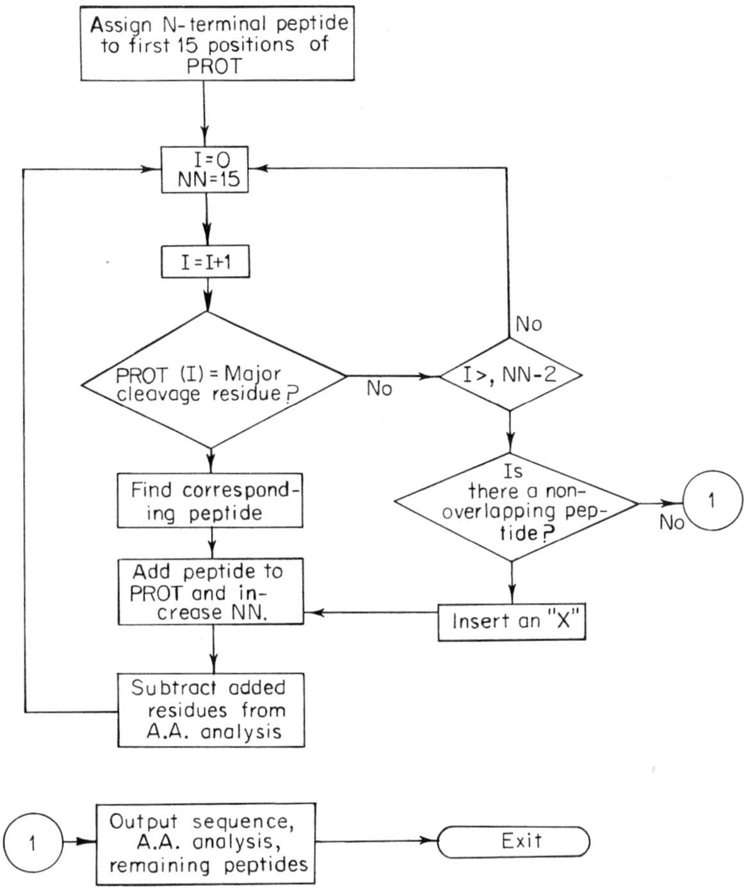

FIG. 11. Flow chart of computer program designed to reassemble protein sequences from N-terminal sequence data from peptide mixtures.

lys-arg bond. Using such cross correlations to compare the four sets of data, the N-terminal sequences of the individual peptides in each set can be deduced.

Because of the complexity of the process, a computer program has been designed to accomplish this feat [71]. The program searches one data set (matrix) for the presence of an element (amino acid) from another data set (matrix). If one is found, then the two matrices are compared using the strategy outlined. Each possible sequence thus obtained is checked against the remaining matrices for authentication, after which it is stored. Once all possible sequences are obtained, the sequence corresponding to the N-terminal sequence is removed and the remaining sequences are used for further overlap. In cases where the information in the N-terminal 15 amino acids is not sufficient to unambiguously define the N-terminal sequence of each peptide in the mixtures, the stepwise degradation can be extended to 20 or 25 residues and/or a fifth or alternative method of cleaving the protein can be used to provide the needed structural information.

Once the N-terminal sequences of the individual peptides are established, this information, together with the N-terminal sequence and amino acid composition of the protein is used to reconstruct the primary structure of the protein by computer analysis of the data. Figure 11 contains a flow diagram of the program. The major steps of the computer program designed for this purpose are as follows:

Step 1 The N-terminal sequence (15-20 residues) forms the initial partial sequence and is searched for a major cleavage residue (i.e., met, cys, arg, or lys).

Step 2 When a major cleavage residue is found, the peptide array is searched for the corresponding peptide. The contributing portion of the peptide is added to the partial sequence and the peptide is removed from the array.

Step 3 Search of the partial sequence for another major cleavage residue is continued.

As each residue is added to the sequence, it is also subtracted from the amino acid analysis data. If at some point the search procedure should terminate (reach the end of the partial sequence) before finding a major cleavage residue, an X is placed in that position and the peptide array is searched for a nonoverlapping peptide (a peptide whose N-terminal sequence does not overlap

```
                                        END OF N-TERMINAL OVERLAP

SEQUENCE IS

     K-V-F-G-R-C-E-L-A-A-A-M-K-R-H-G-L-D-N-R-Y-G-Y-S-L
     G-N-W-V-C-A-A-K-F-E-S-N-F-N-T-Q-A-T-N-R-N-T-D-G-S
     T-D-Y-G-I-L-Q-I-N-S-X-C-A-K-K-I-V-S-D-G-D-G-M-N-A
     W-V-A-W-R-N-R-C-K-G-T-D-V-Q-A-W-I-R-G-C-R-L-X-R-W
     W-C-N-D-G-R-T-P-G-S-R-N-L-C-N-I-P-C-S-A-L-L-S-S-D
     I-T-A-S-V-N-C

                         THE FOLLOWING AMINO ACIDS WERE NOT ACCOUNTED FOR

                                        A     0
                                        F     0
                                        K     0
                                        Y     0
                                        S     0
                                        C    -1
                                        G     0
                                        L     0
                                        P     0
                                        T     0
                                        D     0
                                        H     0
                                        M     0
                                        Q     0
                                        V     0
                                        E     0
                                        I     0
                                        N     0
                                        R     0
                                        W     0

                         (COMPUTER OUTPUT FOR SEQUENCE OF CHICKEN LYSOZYME)
```

FIG. 12. Computer output of reconstructed amino acid sequence of lysozyme.

with any other peptide in the array). This peptide is inserted into the position immediately following the X, and the search for major cleavage residues continues. The program terminates upon either exhausting the peptide array or failing to find a nonoverlapping peptide following the insertion of an X. The latter is extremely rare, since it would require that two peptides have identical sequences in the N-terminal five residues.

Figure 12 shows a simulated computer printout of the reconstructed amino acid sequence of lysozyme together with the tally of amino acids used. The program reversed the two polypeptide fragments beginning at the first X and at the second X. The addition of the C-terminal sequence of the protein of -cys-arg-leu would resolve the ambiguity. In those cases where incomplete sequences are obtained, the data can be expanded. For example, the N-terminal sequence analysis of the peptides can be extended to 25-30 amino acids, especially when the stepwise sequence analysis is performed with an automatic sequencer and/or additional peptide mixtures can be generated using additional enzymatic or chemical fragmentation methods so that the amount of overlapping sequence information is increased.

In the final analysis, while it is unlikely that one method will provide the ultimate solution to the analysis of protein structure and while other strategies may exist to accomplish this goal, the above stragety provides a compromise between the amount of work necessary to generate the sequence data and the degree of reliability with which the reconstructed primary structure can be assembled. Without the use of mass spectrometry many of these protocols are not feasible at all.

In the future it may be that the major contribution of mass spectrometry to the elucidation of protein structure will not reside primarily in its ability to rapidly sequence the small peptides and identify the products of the stepwise Edman reaction, but in its ability to provide data such as those described in this section, so that strategies which reduce the number of man-hours of work and manipulation of the protein necessary to establish the complete primary structure of the protein can be developed and refined.

ACKNOWLEDGMENT

We thank Charles L. Smith for editorial assistance. This research was supported in part by National Science Foundation Grant No. PCM 77-05586.

REFERENCES

1. K. Biemann and W. Vetter, *Biochem. Biophys. Res. Commun.*, *3*, 578 (1960).
2. K. Biemann, F. Gapp, and J. Seibl, *J. Amer. Chem. Soc.*, *81*, 2274 (1959).
3. K. Biemann, *Chimia*, *14*, 393 (1960).
4. R. T. Aplin and J. H. Jones, *Chem. Comm.*, 261 (1967).
5. F. Weygand, A. Prox, H. H. Fassel, and K. K. Sum, *Zeitschift Naturforsch.*, *20b*, 1169 (1965).
6. I. Lengyel, R. A. Salomone, and K. Biemann, *Organic Mass Spectrometry*, *3*, 789 (1970).
7. K. Heyns and H. F. Grutzmacher, *Ann. Chem.*, *669*, 183 (1963).
8. H. R. Morris, D. H. Williams, and R. P. Ambler, *Biochem. J.*, *125*, 189 (1971).
9. M. Barber, P. Jolles, E. Vilkas, and E. Lederer, *Biochem. Biophys. Res. Commun.*, *18*, 469 (1965).
10. D. W. Thomas, B. C. Das, S. P. Gero, and E. Lederer, *Biochem. Biophys. Res. Commun.*, *32*, 519 (1968).
11. E. Lederer, *Pure Appl. Chem.*, *17*, 489 (1968).
12. J. Penders, H. Copier, W. Heerman, G. Dijkstra, and J. F. Crews, *Receuil*, *85*, 216 (1966).
13. M. H. Studier and L. P. Moore, *Biochem. Biophys. Res. Commun.*, *40*, 894 (1970).
14. N. Seiler, H. H. Schneider, and K. D. Sonnenberg, *Anal. Biochem.*, *44*, 451 (1971).
15. G. Marino and V. Buonocore, *Biochem. J.*, *110*, 603 (1968).
16. P. Edman, *Acta Chem. Scand.*, *4*, 283 (1950).
17. H. Tschesche, M. Schneider, and E. Wachter, *FEBS Lett.*, *23*, 367 (1972).
18. F. Weygand and R. Obermeier, *Eur. J. Biochem.*, *20*, 72 (1971).
19. T. Fairwell and R. E. Lovins, *Biochem. Biophys. Res. Commun.*, *43*, 1280 (1971).
20. T. Fairwell, S. Ellis, and R. E. Lovins, *Anal. Biochem.*, *53*, 115 (1973).
21. H. M. Fales, Y. Nagai, G. W. A. Milne, H. B. Brewer, T. Bronzart, and J. J. Pisano, *Anal. Biochem.*, *4*, 288 (1971).

22. H. Tschesche and E. Wachter, *Eur. J. Biochem., 16*, 187 (1970).

23. T. Sun and R. E. Lovins, *Anal. Biochem., 45*, 176 (1972).

24. D. M. Hagins, J. Henke, and R. E. Lovins, *Immunochemistry, 14*, 697 (1977).

25. H. Hagenmaier, W. Ebbighausen, G. Nicholson, and W. Votsch, *Z. Naturforsch., 25b*, 681 (1970).

26. R. E. Lovins, J. Craig, T. Fairwell, and C. McKinney, *Anal. Biochem., 47*, 539 (1972).

27. F. Sanger, *Biochem. J., 39*, 507 (1945).

28. B. S. Hartley and V. Massey, *Biochim. Biophys. Acta, 21*, 58 (1956).

29. R. R. Porter and F. Sanger, *Biochem. J., 42*, 287 (1948).

30. A. L. Levy, *Nature, 174*, 126 (1954).

31. M. Brenner, A. Niederweiser, and G. Pataki, *Experientia, 17*, 145 (1961).

32. W. R. Gray, in *Methods in Enzymology* (C. H. W. Hirs, Ed.), Academic, New York, 1967, Vol. Xi, p. 139.

33. R. Schwyzer and H. Kappeler, *Helv. Chim. Acta, 42*, 2622 (1959).

34. P. Edman and G. Begg, *Europ. J. Biochem., 1*, 80 (1967).

35. J. Pisano and T. Bronzert, *J. Biol. Chem., 244*, 5597 (1969).

36. K. T. Wang, I. S. Y. Wang, A. L. Lin, and C. S. Wang, *J. Chromat., 26*, 323 (1967).

37. H. Van Order and F. Carpenter, *Biochem. Biophys. Res. Commun., 14*, 399 (1964).

38. V. M. Stepanov and V. F. Kristsov, *J. Gen. Chem. (USSR), 35*, 556 (1965).

39. B. W. Melvas, *Acta Chem. Scand., 23*, 1679 (1969).

40. F. F. Richards and R. E. Lovins, in *Methods in Enzymology* (C. H. W. Hirs, Ed.), Academic, New York, 1972, Vol. XXV, p. 314.

41. A. W. Brauer, M. N. Margolis, and E. Haber, *Biochemistry, 14*, 3029 (1975).

42. J. M. Brewer, T. Fairwell, J. Travis, and R. E. Lovins, *Biochemistry, 9*, 1011 (1970).

43. A. P. Graffeo, G. Hoag, and B. L. Karger, *Anal. Lett., 6*, 505 (1973).

44. C. L. Zimmerman, J. J. Pisano, and E. Apella, *Biochem. Biophys. Res. Commun., 55*, 1220 (1973).

45. W. R. Gray and G. del Valle, *Biochemistry, 9*, 2134 (1970).

46. R. Kuhn, I. Low, and H. Frischmann, *Ber., 90*, 203 (1957).

47. J. Coggins and L. Benoitin, *Biol. Abstr., 18*. 156th ACS National Meeting, Atlantic City, Sept. 1968.

48. P. A. Leclercq and D. M. Desiderio, *Biochem. Biophys. Res. Commun., 45*, 308 (1971).

49. M. L. Polan, W. J. McMurray, S. R. Lipsky, and S. Lande, *J. Amer. Chem. Soc., 94*, 2847 (1972).

50. M. M. Shermyakin, Y. A. Ovchinnikov, E. I. Vinogradova, M. Y. Feigina, A. A. Kiryushkin, N. A. Aldanova, B. Alakhov, M. V. Lipkin, and B. V. Rosinov, *Experientia, 23*, 428 (1967).

51. J. Lenard and P. M. Gallop, *Anal. Biochem., 29*, 203 (1969).

52. G. R. Stark, *Biochemistry, 7*, 1796 (1968).

53. P. Schlack and W. Kumpf, *Hoppe-Seylars Z. Physiol. Chem., 154*, 125 (1926).

54. M. Rangarojan, R. E. Ardery, and A. Darbre, *J. Chromat., 87*, 499 (1973).

55. R. F. Goldberger and C. B. Anfinsen, *Biochemistry, 1*, 401 (1962).

56. P. J. Butler, J. I. Harris, B. S. Hartley, and R. Leberman, *J. Biochem., 103*, 78 (1967).

57. L. J. Slobin and S. J. Singer, *J. Biol. Chem., 243*, 1777 (1968).

58. E. Gross and B. Witkop, *J. Amer. Chem. Soc., 83*, 1510 (1961).

59. L. K. Ramachandran and B. Witkop, *Meth. Enzymol., 11*, 283 (1967).

60. C. M. Isung and H. Fraenkel-Conrat, *Biochemistry, 4*, 793 (1965).

61. J. K. McDonald, S. Ellis, and T. J. Reilly, *J. Biol. Chem., 241*, 1494 (1966).

62. R. M. Caprioli and W. E. Seifert, *Biochem. Biophys. Res. Commun., 64*, 295 (1975).

63. T. Mahowald, *Biochemistry, 4*, 732 (1965).

64. A. M. Katz, W. J. Dreyer, and C. B. Anfinsen, *J. Biol. Chem., 234*, 2897 (1959).

65. W. A. Schroeder, R. T. Jones, J. Cormick, and K. McCalla, *Anal. Chem., 34*, 1570 (1962).

66. H. Nau, *Biochem. Biophys. Res. Commun., 59*, 1088 (1974).

67. V. Yu. Garilov, A. D. Frank-Kamenetskii, and M. D. Frank-Kamenetskii, *Biokhimiya, 31*, 799 (1966).

68. G. Braunitzer, R. Chen, B. Schrank, and S. Stange, *Hoppe-Seylers Z. Physiol. Chem., 353*, 832 (1972).

69. W. R. Gray, *Nature, 220*, 1300 (1968).

70. C. Perrine, E. L. Cannon, C. Cavender, T. Fairwell, and R. E. Lovins, *Adv. Mass Spectr., 5*, 728 (1971).

71. E. L. Cannon and R. E. Lovins, *Anal. Biochem., 46*, 33 (1972)

Chapter 3

FOOD AND FLAVOR INDUSTRY

MICHAEL G. KOLOR

Central Research Laboratories
General Foods Technical Center
Tarrytown, New York

I. INTRODUCTION

Mass spectrometry has provided the flavor chemist with an objective means of identifying the chemical components which give each food its unique flavor qualities. Identification of flavor compounds in foods by classical chemical techniques is extremely difficult, if not impossible, because of the complex nature of food systems and the minute levels at which these chemicals exist in the food. One of the first reported identifications of an important flavor component was accomplished by Kleber [1] in 1913 when he identified amyl acetate in banana by laborious classical techniques. While this was a major advance in its day, it has become a relatively simple identification through the use of mass spectrometry.

During the late 1940s the petroleum chemist began to use the relatively new technique of mass spectrometry as an analytical tool to quantitatively and qualitatively study the volatile components of petroleum fractions. The organic chemist began to realize the mass spectrum of a chemical provided a fingerprint of that chemical and, more importantly, it could be used for structure elucidation. The food chemist recognized soon after that mass spectrometry was an extremely powerful tool for unraveling the complex nature of food flavor systems.

The marriage of gas chromatography (GC) and mass spectrometry (MS) provided the food analyst with a revolutionary analytical technique. With this ideal combination it became possible to separate and identify chemical components from a complex flavor isolate even though the component of interest was present at a very low concentration. It is not surprising then that flavor analysis, where high degrees of sensitivity and resolution are prime requisites, was one of the fields of research which played a major role in pioneering the development of tandem chromatography-mass spectrometry analysis. The mass spectrometer has become an invaluable tool in the food and flavor research laboratory. Over the past 15 years the literature has been flooded with papers describing the compositional make up of food flavors as a direct result of mass spectral analysis.

Although primarily utilized by the food industry for the identification of volatile flavor components, mass spectrometry is finding more and more use as an analytical technique in analyzing the nonvolatile food components which make up the bulk of the food product, contribute to nutrition, basic taste, and often serve as precursors to the more volatile flavor compounds. Derivatization techniques are commonly employed to increase the volatility of sugars, amino acids, peptides, and lipids, thereby making these relatively nonvolatile compounds amenable to analysis by mass spectrometry.

II. TECHNIQUES USED IN FLAVOR ANALYSIS

A. Isolation and Separation

Before the mass spectrometer can become useful in flavor analysis, the flavor essence must first be isolated from a food system and separated into its chemical components. Figure 1 represents a simplified diagram of the typical procedures used for the analytical study of food flavors. Isolation of the volatile flavor components from a food system is usually achieved by distillation and solvent extraction used alone or in combination. Vacuum distillation in an inert atmosphere is preferable, especially for fresh uncooked foods such as fruits, since the

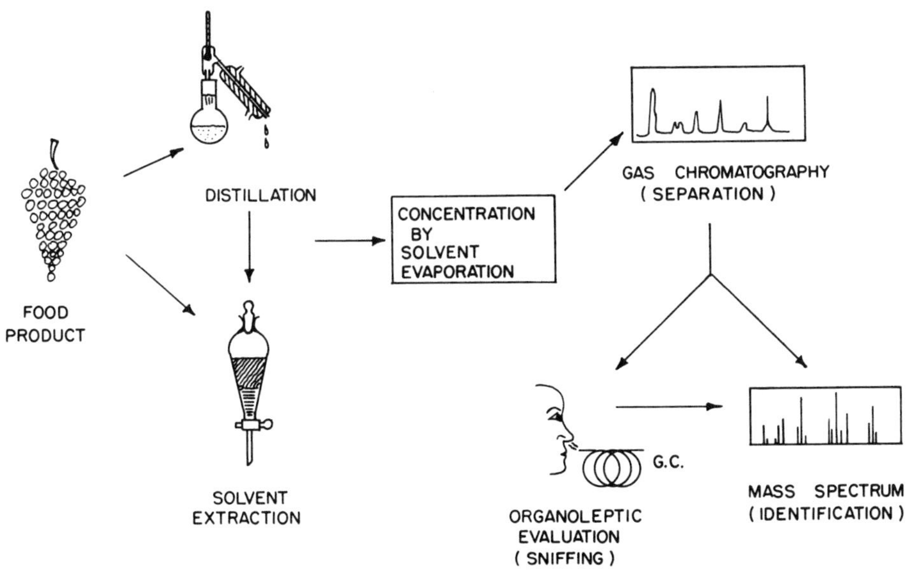

FIG. 1. Flow diagram of food flavor analysis.

possibility of artifact formation and/or losses of labile chemicals by heat degradation or oxidation is kept to a minimum. The flavor components are then usually extracted from the aqueous distillate with a volatile solvent such as ether which can be evaporated under a stream of inert gas, thereby concentrating the flavor components in the solvent.

Concentration of the flavor essence is necessary since the total amount of flavor components in a food product is usually only a small fraction of a percent. Many chemicals play a very important role in the flavor impact of a food at or below the parts per billion (ppb) range. The extreme sensitivity of the nose is well illustrated by 2-methoxy-3-isobutylpyrazine which is primarily responsible for the flavor of green peppers and can be detected at a concentration of 0.002 ppb [2].

The concentrated flavor isolate, which may contain hundreds of compounds over a wide range of concentrations, must now be separated into its individual components. Gas chromatography is ideally suited for this challenge because of its high sensitivity and high resolving power. Once these compounds have been separated, mass spectrometry becomes an indispensible tool for structure elucidation and ultimately their identification. The technique of combined gas chromatography-mass spectrometry and its application in organic analysis has been throughly discussed [3].

B. Organoleptic Evaluation

In some food systems a single chemical is primarily responsible for the characteristic flavor, with other compounds contributing subtle notes to the overall flavor impact. In other foods, however, a blend of chemicals at precise levels appears to be responsible for the flavor properties. Evaluation of a gas chromatographic effluent by means of smelling the emerging components provides a very important method of determining the contribution

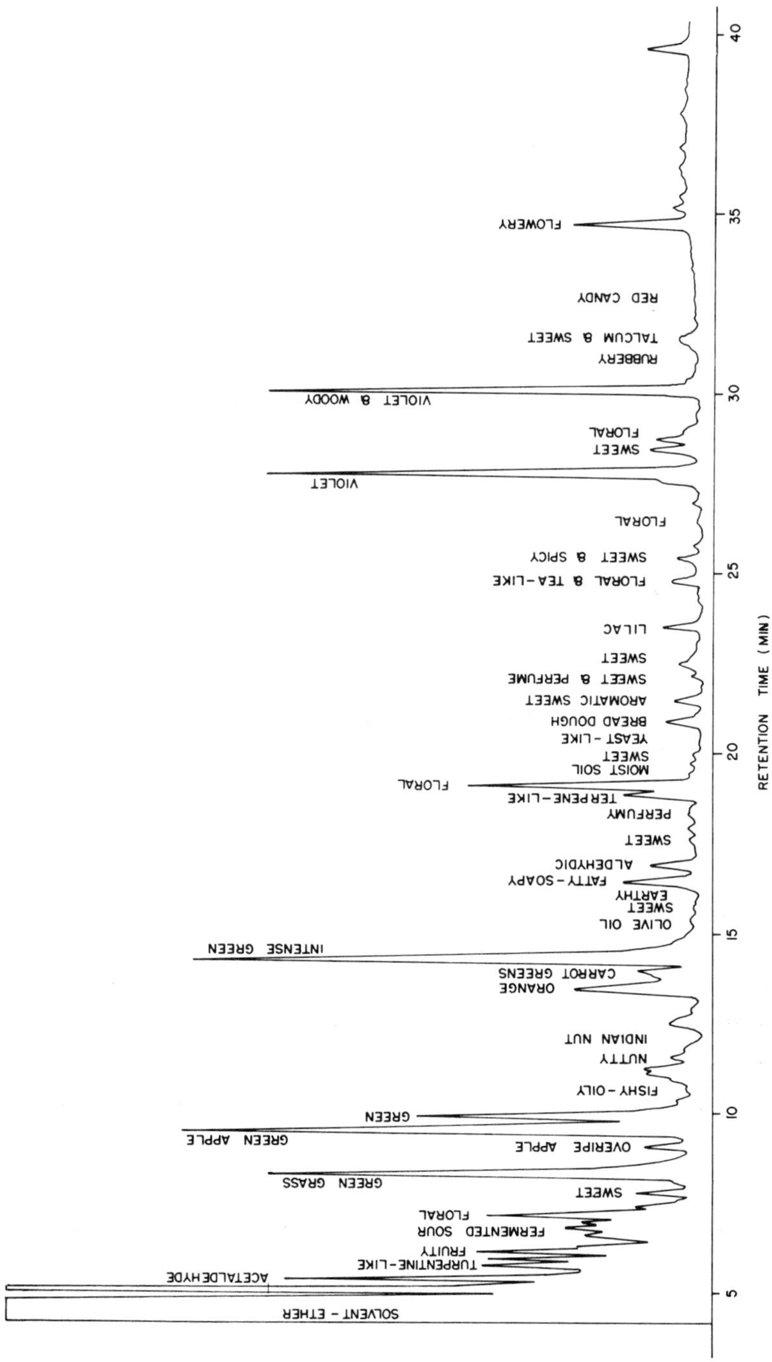

FIG. 2. Organoleptic evaluation of red raspberry flavor isolate chromatographed on an 8 ft x 1/8 in. SP-1000 column. Programmed from 70°C to 250°C at 6°/min.

of the various chemicals to the overall flavor of the food. This subjective evaluation of flavor chemicals is an art in itself and becomes a very powerful approach to flavor analysis in the hands of a skilled flavorist. A compilation of literature data on flavor and taste threshold values has been published which tabulates the detection and recognition threshold values of 1238 flavor compounds [4].

A typical organoleptic evaluation ("sniffing") of a fruit flavor essence is shown in Figure 2. The volatile flavor distillate from fresh red raspberries was passed through a gas chromatograph and the separated components were evaluated by smelling the eluant from the gas chromatograph. The effluent was split so that a portion was directed into a mass spectrometer where the total ion current was monitored to produce a total ion chromatogram of the raspberry components. In this way the flavor sample is analyzed by simultaneously monitoring: the quantitative distribution of components by means of the total ionization in the mass spectrometer, the aroma properties of the components by means of "sniffing" the emerging components, and the identification of these components from their mass spectra.

A wide variety of aromas are encountered in the organoleptic evaluation of the raspberry flavor essence. A still greater variety and complexity of aromas are encountered in heat-treated and processed foods such as meat and coffee flavor isolates. Sometimes the skilled flavorist can identify, from aroma alone, the chemicals emerging from the gas chromatograph. However, if the flavorist has not previously encountered a particular chemical, the mass spectrum becomes the key to its identification.

C. Headspace Analysis

The high sensitivity of tandem gas chromatography-mass spectrometry makes possible the direct sampling and analysis of the

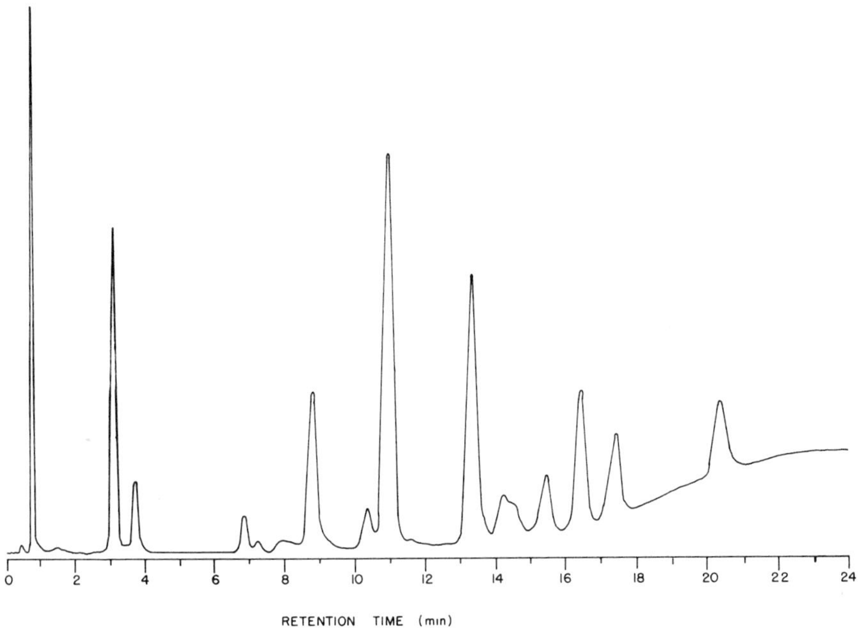

FIG. 3. Gas chromatogram of a coffee headspace sample from a 6 ft x 1/8 in. Porapak-QS column. Programmed from 70°C to 200°C at 8°/min.

volatile components in the headspace over a food product. While this method is primarily useful for analysis of the most volatile compounds, techniques have been developed for concentrating the relatively less volatile chemicals on porous polymers and other reversibly adsorbant materials [5,6]. This technique requires a minimum of sample handling which reduces the chance of artifact formation and losses of highly volatile or labile components. Figure 3 represents the gas chromatogram of the headspace obtained by means of a gas-tight syringe directly (without a concentration step) from a can of roasted coffee.

Table 1 lists the major components in the coffee headspace sample which were identified from their mass spectra.

TABLE 1

Components of Coffee Headspace

Retention Time (min)	Identity
0.7	Methane
3.1	Ethylene
3.8	Ethane
6.8	Propene
8.8	Acetaldehyde
10.4	Methyl formate
11.0	Butene
13.4	Acetone + furan
14.3	Dimethyl sulfide
15.5	Isoprene
16.5	Isobutyraldehyde
17.5	Diacetyl + methylfuran
20.4	Isovaleraldehyde

D. Specific Examples

The identification of trans-2-nonenal as a compound which imparts a fresh-brewed "woody" character to roasted and ground coffee [7] serves as an excellent example of the various techniques that can be employed to identify a component at very low concentration in a very complex mixture. The complexity of the gas chromatographic pattern from coffee flavor fractions hindered earlier efforts to characterize and identify this very small component. It was noticed, however, during the organoleptic evaluation of a green pepper flavor isolate that one area of the gas chromatogram possessed this same "woody" note. Even though the green pepper flavor fraction contains many components, it is not nearly as complex as a coffee flavor isolate where, besides the naturally occurring compounds, many new compounds are produced by browning reactions during the roasting and processing of the coffee beans. Approximately 400 compounds have been identified in coffee flavor [8]. An example of the complexity of a coffee vacuum steam distillate is illustrated

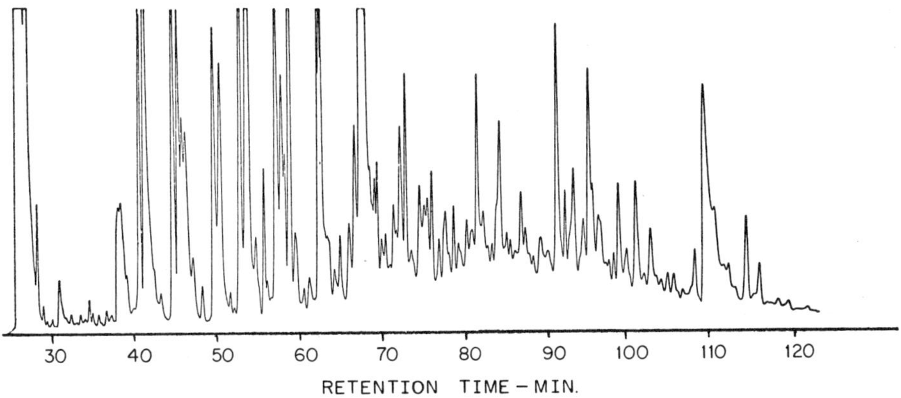

FIG. 4. Gas chromatogram of a coffee vacuum steam distillate from a 500 ft x 0.03 in. (inside diameter) capillary column coated with Carbowax-20M. Programmed from 70°C to 200°C at 1.5°/min.

by the gas chromatogram obtained on a 500 ft capillary column (Fig. 4). Therefore, an effort was made to identify this compound from the bell pepper distillate.

Mass spectra were obtained of the compounds in the area of the gas chromatogram where the woody note was found by means of tandem gas chromatographic-mass spectrometric analysis. Unfortunately, none of the major compounds identified in this area of the chromatogram possessed the sought-after aroma. From these results, the assumption was made that the woody component was a very dilute compound in this area of the chromatogram and that its intense aroma overshadowed the aromas of the major constituents. Attempts to separate this compound by gas chromatography of the bell pepper distillate on different liquid phases were also unsuccessful.

Because of the importance of this flavor note, it was decided to scale up the entire isolation procedure in order to collect a sufficient sample by repeated trapping of the gas chromatographic effluent so that it could be rechromatographed on as many successive liquid phases as were required to effect

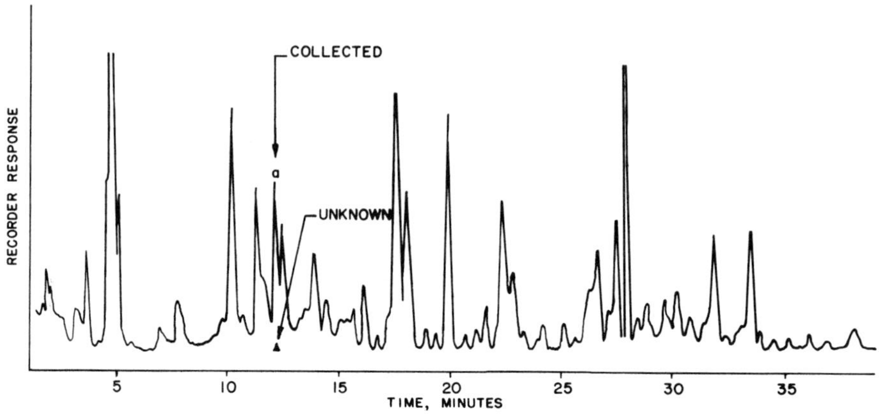

FIG. 5. Gas chromatogram of a green pepper flavor isolate from a 6 ft x 1/4 in. 20% SE-52 column. Run isothermally at 150°C.

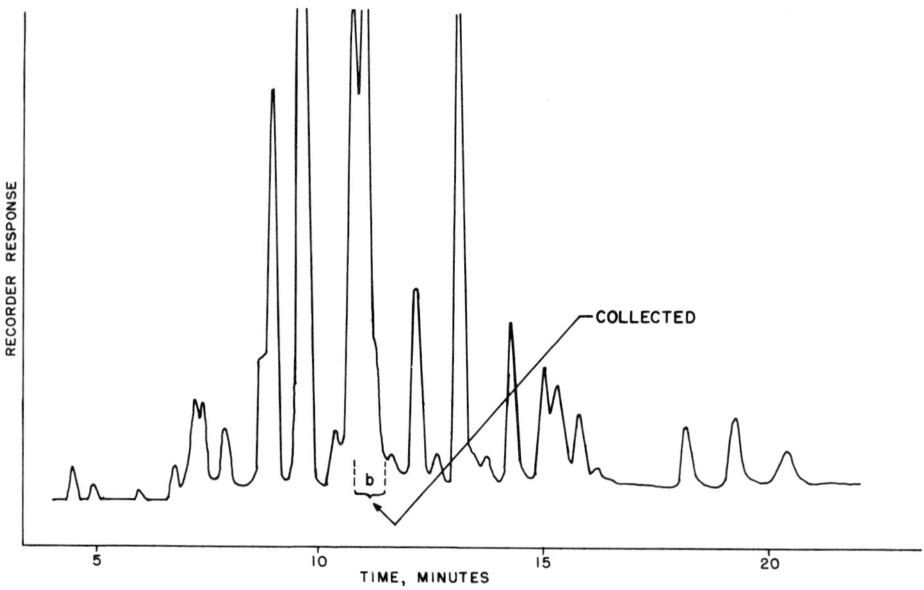

FIG. 6. Area a (from Fig. 5) collected and rechromatographed on a 10 ft x 1/4 in. 10% DEGS column. Run isothermally at 130°C.

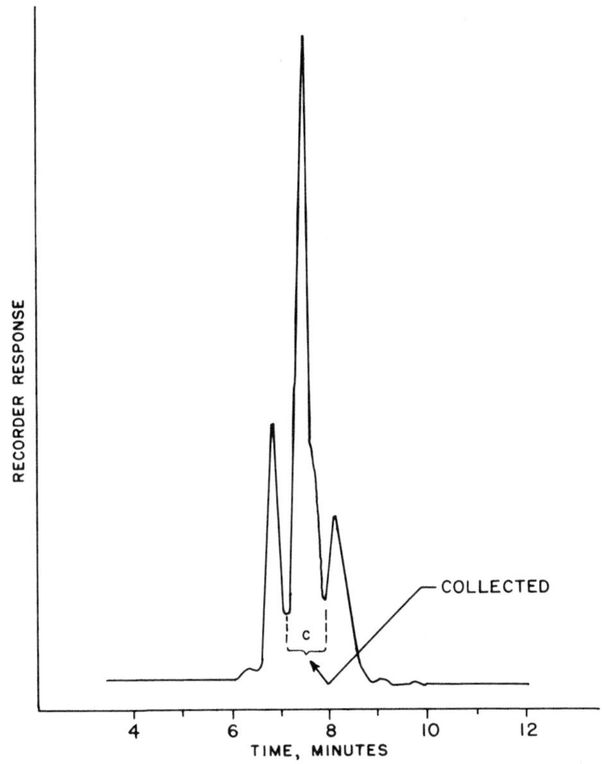

FIG. 7. Area b (from Fig. 6) collected and rechromatographed on a 6 ft x 1/8 in. 10% Carbowax-20M column. Programmed from 100°C to 200°C at 6°/min.

complete separation. Figures 5 to 8 show the original and three successive gas chromatograms, obtained in each case by repeated trapping of the effluent and reinjection on a different liquid phase until a pure sample of the woody compound was obtained. Throughout the purification, organoleptic evaluation of the effluent from each of the liquid phases was used to locate the desired component. The location of the woody aroma is represented by a, b, c, and d in the gas chromatograms. Sufficient pure sample was collected from the final liquid

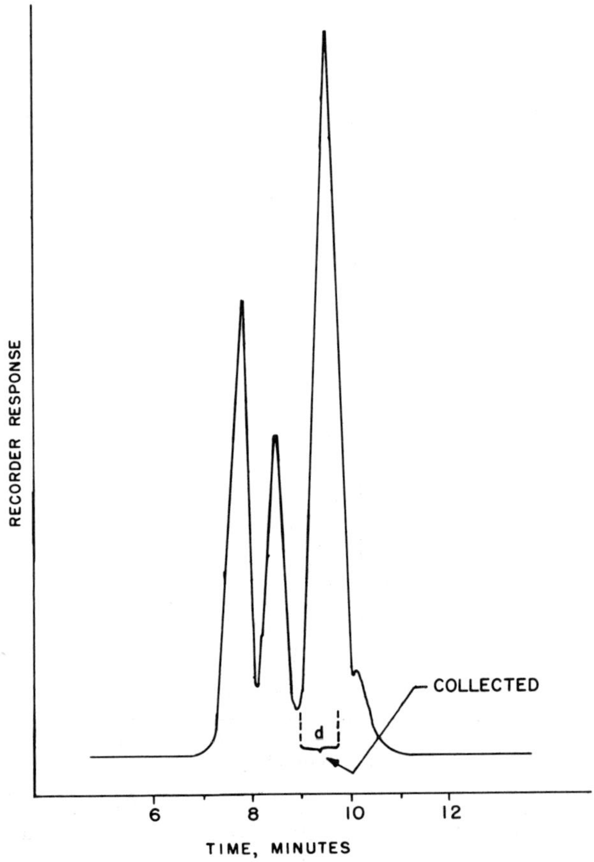

FIG. 8. Area c (from Fig. 7) collected and rechromatographed on a 6 ft x 1/8 in. 15% OV-225 column. Programmed from 100°C to 200°C at 6°/min.

phase so that high-resolution mass spectrometry, infrared spectrometry, and nuclear magnetic resonance (NMR) spectrometry could be employed, if necessary, to give a complete structural identification of this compound.

The mass spectrum of this compound is given in Figure 9. High-resolution analysis showed that the molecular ion at m/e 140 had an empirical composition of $C_9H_{18}O$. The intense ions

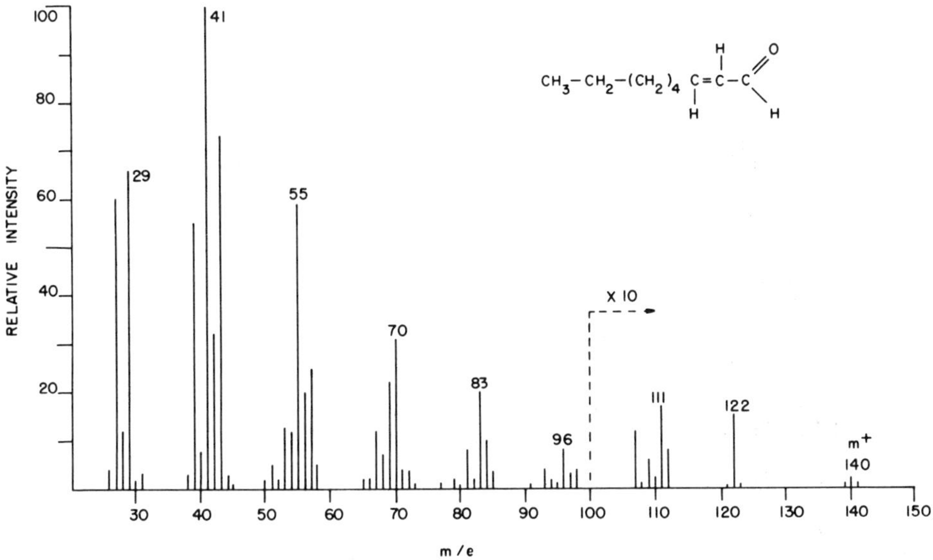

FIG. 9. Mass spectrum of 2-nonenal: CEC 21-110, 70 eV, heated glass inlet system.

at m/e 41 ($C_3H_5^+$) and m/e 55 ($C_4H_7^+$) indicated unsaturation. The series of ions at m/e 139, 122, 112, and 96 representing losses of 1, 18, 28, and 44 from the molecular ion is characteristic of aldehydes. From these data a nine-carbon unsaturated aldehyde was suspected. The infrared spectrum indicated that the compound was an α,β-unsaturated aldehyde and that there was trans configuration about the double bond. Nuclear magnetic resonance data were in agreement with these data. This example shows the need in some cases to have supplementary spectral data in order to obtain a complete structural identification.

The identification of trans-2-nonenal as the woody compound was at first hard to believe because this chemical was known to possess a strong cucumber-like, fatty and rancid oil aroma. However, at dilutions of 0.4 to 2 ppb this chemical was indeed found to have the sought-after woody aroma and taste. It is well known that the aroma of many chemicals changes dramatically at different dilutions. The above procedure was repeated on a large quantity of coffee beans, and the presence of trans-2-nonenal as the woody note in fresh-brewed coffee was proven. An estimate of the small quantity of this compound in the original bell pepper distillate is shown by the small dark triangle in Figure 5. This analysis demonstrates the typical procedures and some of the innovations that sometimes must be developed for the characterization and identification of important flavor compounds which are very dilute in extremely complex mixtures.

III. MASS SPECTRA OF VOLATILE FOOD FLAVOR COMPOUNDS

A comprehensive listing of the volatile compounds identified in 90 food products is given in a recent report by the Netherlands Central Institute for Nutrition and Food Research [8]. A total of 2132 different chemical compounds from 914 references were reported in these 90 foods. A broad spectrum of chemical structures containing carbon, hydrogen, oxygen, nitrogen, and sulfur have been identified in the flavor isolates of these foods. Esters appear to be the class of compounds most abundantly found in food products, with a total of 487 being reported. Carbonyl compounds are the next most abundantly distributed class of compounds, with 184 ketones and 122 aldehydes being identified. Other classes of compounds found in these foods include 260 hydrocarbons (including terpenes), 202 alcohols, 137 organic acids, 45 lactones, 163 sulfur-containing compounds, 156 basic compounds (including amines, pyrazines, pyridines, and pyrroles), and 336 compounds classified as miscellaneous, which are comprised of acetals, ethers, nitriles, amides, phenolics, furan compounds, pyrroles, and other heteroatomic and polyfunctional compounds.

The many correlation studies reported in the literature which deal with mass spectral fragmentation modes as related to chemical structure are extremely valuable in the interpretation of the mass spectra produced by the wide variety of chemical structures found in food systems. Reference to these correlation studies and the interpretation of the mass spectra of organic compounds are described in detail in several books [9-16]. The role of mass spectrometry in flavor chemistry has been reviewed [17-19]. A table has been constructed which shows the common fragmentation patterns and rearrangement processes that are encountered in the mass spectra of the main classes of volatile organic compounds found in food systems. [20].

As previously stated, a single chemical is often a primary contributor to the distinctive and unique flavor of a particular food. The diversity in structure of these chemicals provide a wealth of examples for a discussion of the mass spectra of typical flavor compounds. Methyl anthranilate has been identified in grapes [21] and possesses a strong grapelike aroma. The mass spectrum of methyl anthranilate is shown in Figure 10. The

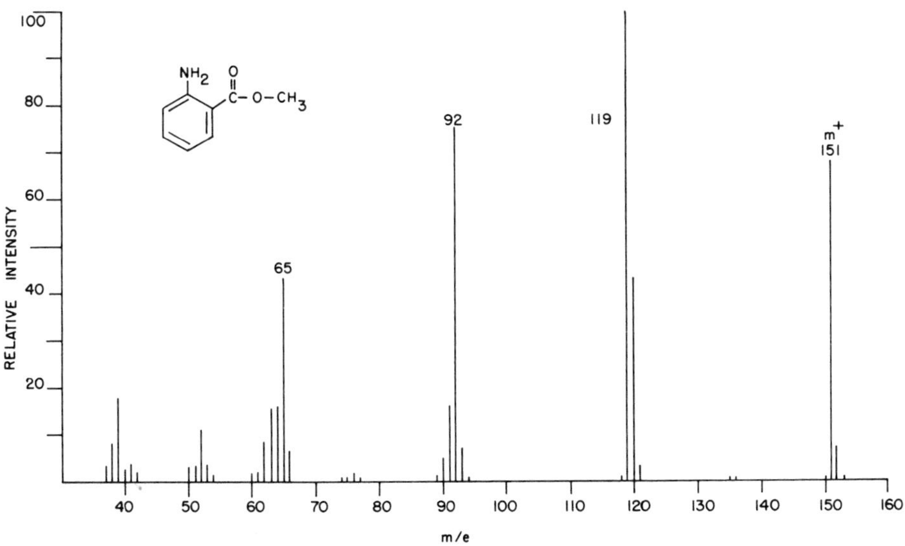

FIG. 10. Mass spectrum of methyl anthranilate: CEC 21-110, 70 eV, heated glass inlet system.

molecular ion is readily observed at m/e 151. The presence of a nitrogen atom in the molecule is obvious from the nitrogen rule which states that the molecular weight of a molecule containing C, H, N, O, and/or S atoms will be an odd number only if there are an uneven number of nitrogen atoms in the molecule. This follows since nitrogen has an odd valency and an even atomic weight, whereas C, O, and S all have even atomic weights and valencies and H has an odd atomic weight and valency. Aromaticity is indicated in this spectrum by the stability of the molecular ion and the intense ions at m/e 39 ($C_3H_3^+$) and m/e 65 ($C_5H_5^+$). The base peak at m/e 119 is shown by a metastable ion at $\dot{m}$* 93.8 to be derived directly from the molecular ion by a rearrangement process involving a loss of 32 mass units from the molecular ion. The loss of methanol from the molecular ion is a common rearrangement process in the spectra of ortho-methyl benzoates. High-resolution analysis of the prominent ion at m/e 92 reveals an empirical formula of $C_6H_6N^+$. This ion represents the loss of 59 mass units from the molecular ion which is a common fragmentation process in methyl esters of carboxylic acids. Taken together, these data strongly suggest that the compound contains a benzene ring substituted by an amino group ortho to a $COOCH_3$ group and the spectrum is in fact methyl anthranilate.

Raspberry ketone is the pseudonym used in the flavor world for 4-(para-hydroxyphenyl)-2-butanone because it has an aroma which is strongly reminiscent of raspberries. This compound was not found in the red raspberry flavor distillate (Fig. 2) even though it had been previously reported in the literature as a component of raspberries [22]. Attempts to chromatograph pure 4-(para-hydroxyphenyl)-2-butanone on the Carbowax 20M column which has an upper temperature limit of 200°C were unsuccessful. It could, however, be chromatographed on OV-101 and SP-1000 liquid phases where the upper temperature limits are extended to 300° and 275°C, respectively. The raspberry flavor distillate was rechromatographed on both OV-101 and SP-1000 columns but once again raspberry ketone was not found. Reexamination of the literature showed that raspberry ketone had not been found by other workers using distillation procedures, but that it had been identified where the flavor was isolated by solvent extraction. Therefore, it was assumed that this high boiling compound was not readily steam-distillable. Direct extraction of crushed red raspberries with ether, concentration of the ether fraction, and analysis by tandem GC-MS using an OV-101 column did, in fact, show that 4-(para-hydroxyphenyl)-2-butanone was a component of red raspberry flavor.

This example demonstrates two places where the analyst must be careful. First of all, using two or more chromatographic liquid phases to analyze a flavor isolate is desirable. Also, while distillation techniques are usually the best means of flavor volatiles isolation, there are times when solvent extraction procedures should be used as auxiliary techniques. When solvent extraction is used as the isolation procedure, the analyst must be wary of artifact formation in the hot injection port of the gas chromatograph from solvent extractable lipids and other nonvolatiles.

The mass spectrum of 4-(para-hydroxyphenyl)-2-butanone is shown in Figure 11. The molecular weight is recognizable as 164. An aromatic ring is suggested by the ions at m/e 77 ($C_6H_5^+$) and 65 ($C_5H_5^+$). The ion at m/e 91 indicates that the benzene ring is substituted by an alkyl group yielding a tropilium ion, and the ion at m/e 94 indicates ring substitution by a hydroxyl group. The base peak in the spectrum at m/e 107 is a characteristic fragmentation process in the spectra of alkyl phenols. The molecular ion at m/e 164 minus the hydroxybenzyl portion of the molecule at m/e 107 leaves a mass of 57, indicating that the entire alkyl group substituent contains five carbons (C_5H_{11}).

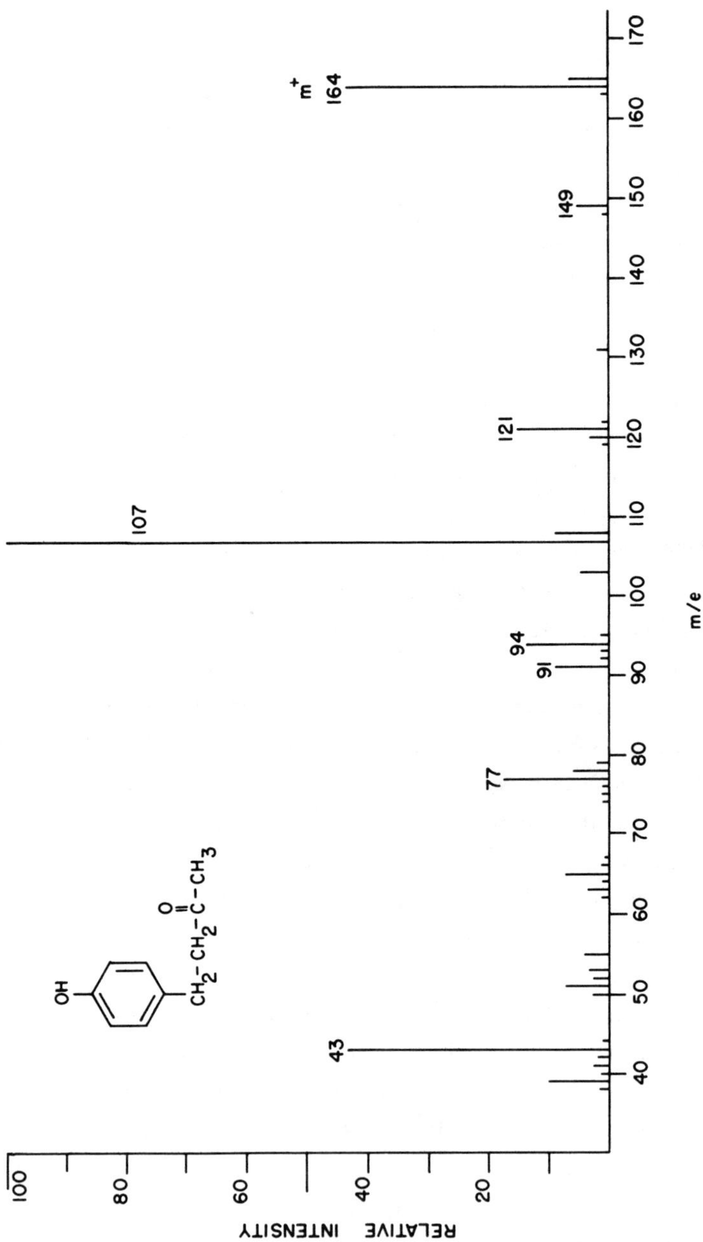

FIG. 11. Mass spectrum of 4-(para-hydroxyphenyl)-2-butanone: DuPont 21-491, 70 eV, GC-MS inlet system.

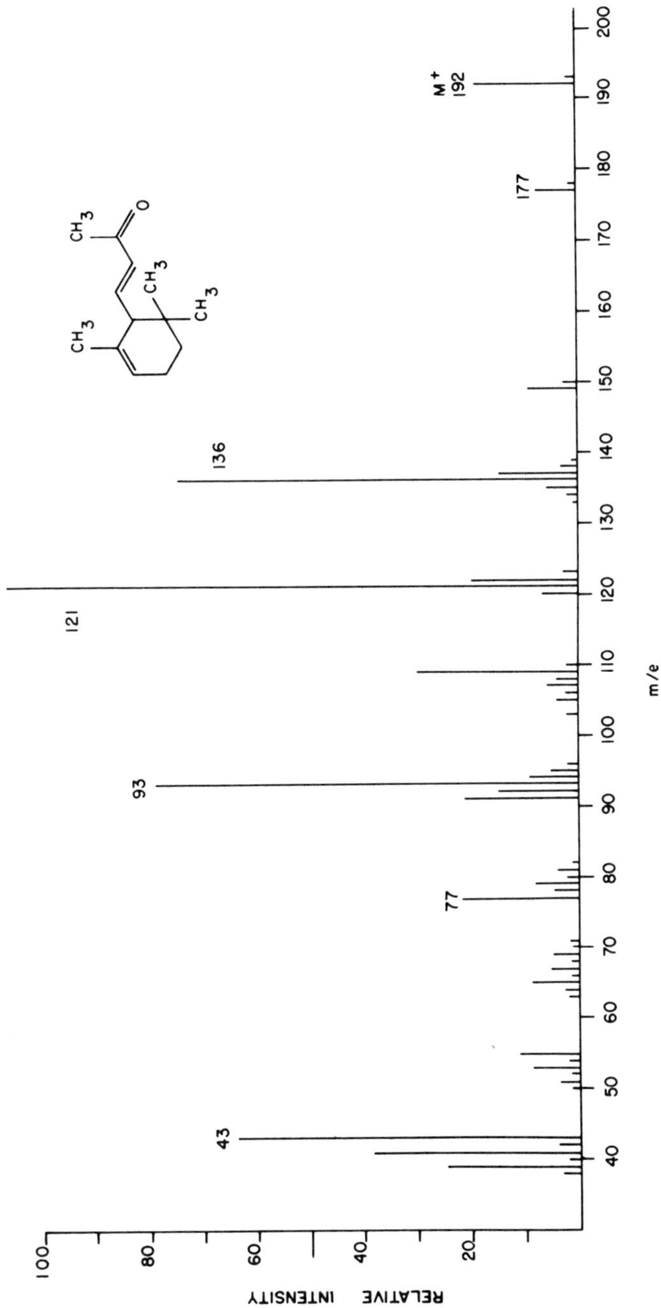

FIG. 12. Mass spectrum of α-inone: DuPont 21-491, 70 eV, GC-MS inlet system.

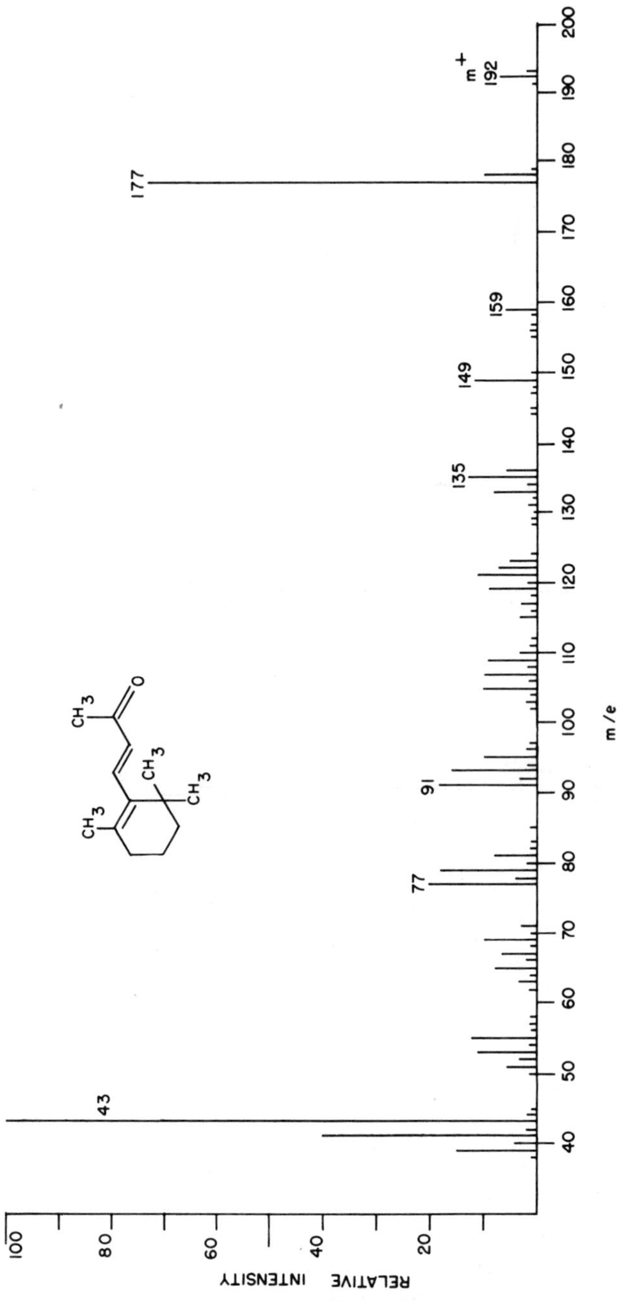

FIG. 13. Mass spectrum of β-ionone: DuPont 21-491, 70 eV, GC-MS inlet system.

Mass measurement of the molecular ion reveals the presence of two oxygen atoms in the molecule. These data together with the absence of an ion at m/e 135 (M - C_2H_5) and the presence of a prominent ion at m/e 121, which is measured as the loss of CH_3CO from the molecular ion, indicate that the suspected pentyl group is actually a 2-butanone group. The position of the substituents on the ring cannot be predicted from the mass spectrum. NMR and GC retention time data can be used to determine that the compound is the para isomer.

α-Ionone and β-ionone are also very important flavor components in raspberry essence [23]. They were evaluated organoleptically (Fig. 2) as having violet and woody-violet fragrances. The mass spectra of α-ionone (Fig. 12) and β-ionone (Fig. 13) represent examples of two isomers with very different mass spectra. Although these two isomers differ only in the position of the double bond, their spectra are markedly different, and they can be differentiated solely from their mass spectra without the need of other spectral data. The fragmentation processes involved in these spectra have been described in detail [24,25]. Cleavage of the methyl group attached to the ring double bond produces the dominant ion at m/e 177 in β-ionone, while cleavage of the allylic carbon-carbon bond in the ring of α-ionone gives rise to the expulsion of an isobutylene molecule yielding an intense ion at m/e 136.

2,5-Dimethyl-4-hydroxy-3(2H)-furanone is an important flavor component of pineapple [26] and strawberries [27]. It is described as having a burnt pineapple, a sweet, and a cotton-candy-like aroma. Analysis by combination GC-MS of a strawberry vacuum-steam distillate in the author's laboratories failed to reveal the presence of this compound. It is known to be a high boiling, unstable compound. Therefore, it was suspected that this labile compound was either not readily steam-distillable or that it did not survive the vacuum steam distillation procedure. Accordingly, fresh strawberries were extracted with ether, and the concentrated ether extract was analyzed by GC-MS during the same day. 2,5-Dimethyl-4-hydroxy-3(2h)-furanone was positively identified as a component of this strawberry extract. This analysis emphasizes the importance of the proper choice of the isolation procedure, and that more than one isolation technique may be needed to obtain a complete profile of the flavor components. Also, in the case of labile compounds GC-MS analysis should be performed as soon as possible after the flavor has been isolated, and care should be taken to prevent reaction or decomposition.

The mass spectrum of 2,5-dimethyl-4-hydroxy-3(2H)-furanone is shown in Figure 14. Identification of the complete structure of this compound, when encountered as an unknown, from the mass spectrum alone is very difficult and other spectral data are required to define the structure. The molecular ion at m/e 128 was found by mass measurement to have an empirical formula of $C_6H_8O_3$. The base peak at m/e 43 was found to be CH_3CO. Other prominent ions at m/e 57 and m/e 85 (M - 43) offer little structural information. The identification of this compound from MS, IR, NMR, and UV spectral data has been described in detail [26]. The IR spectrum of this compound indicates the presence of a hydroxyl group and a conjugated carbonyl group. The NMR spectrum indicates the presence of an OH, a CH, and two CH_3 groups. This example points out that when mass spectral data are backed up by these other spectral techniques the analyst is usually provided with the information necessary to make a complete structural identification of the compound in question.

The mass spectrum of another component in the strawberry vacuum distillate is depicted in Figure 15. This component was organoleptically evaluated in the gas chromatographic effluent and found to possess a strong green aroma with a sweet candylike undernote. It was purified by trapping from the GC column and

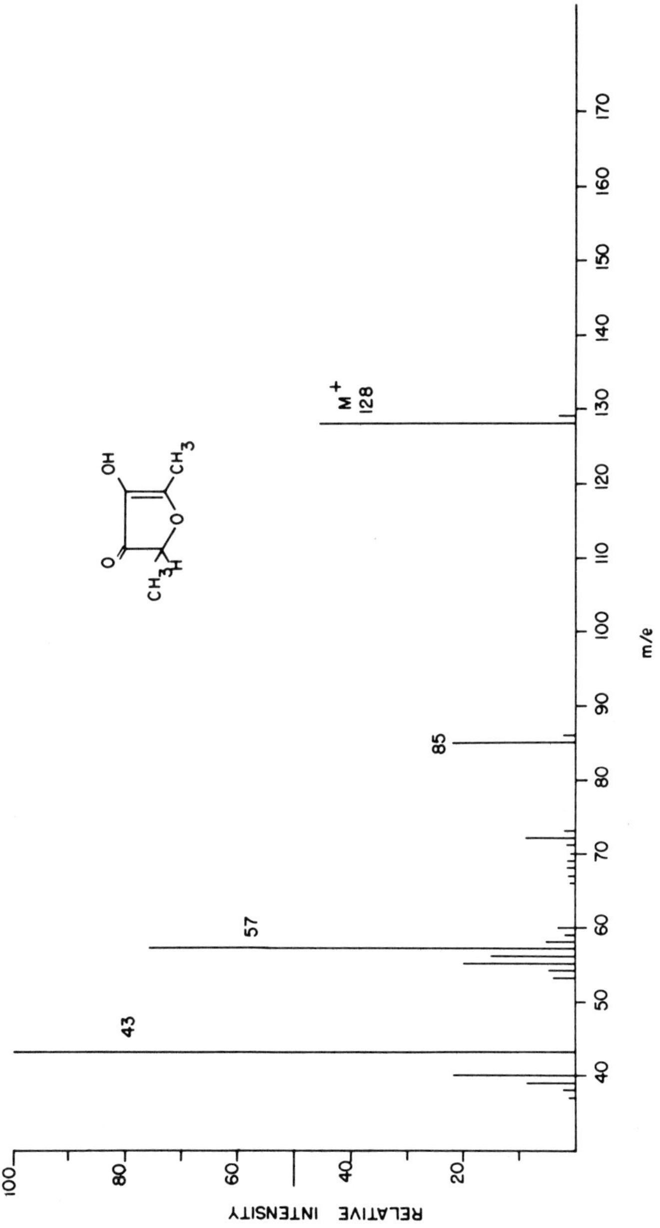

FIG. 14. Mass spectrum of 2,5-dimethyl-4-hydroxy-3(2H)-furanone: DuPont 21-491, 70 eV, GC-MS inlet system.

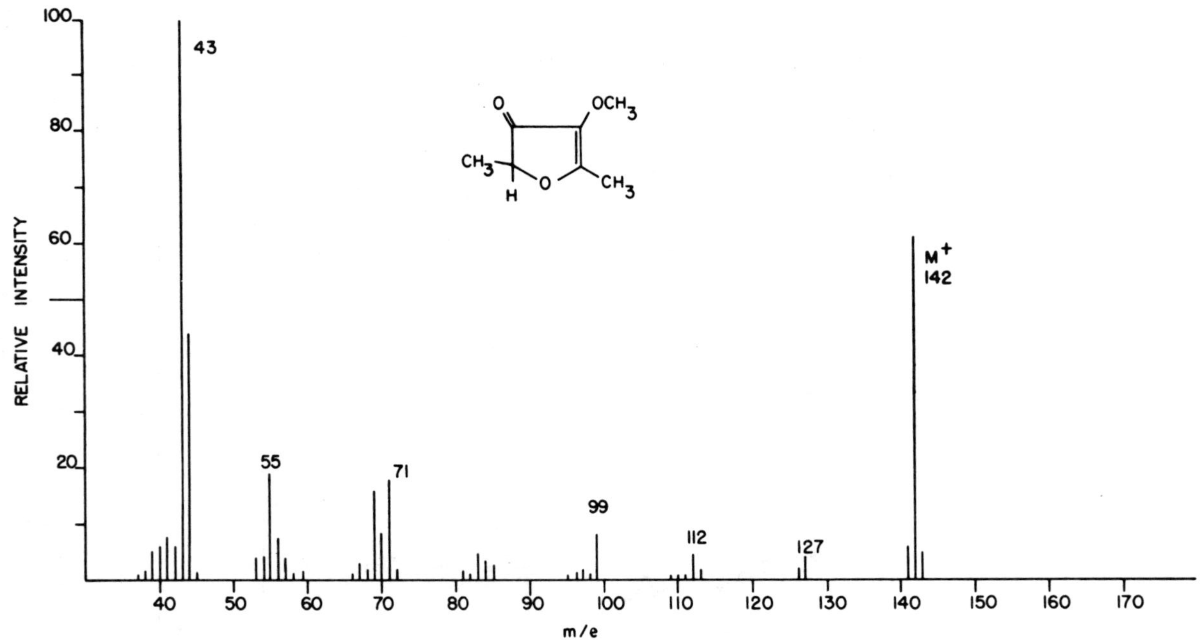

FIG. 15. Mass spectrum of 2,5-dimethyl-4-methoxy-3(2H)-furanone: DuPont 21-491, 70 eV, GC-MS inlet system.

analyzed by high-resolution mass spectrometry. The molecular ion at m/e 142 was mass measured and found to have an empirical formula of $C_7H_{10}O_3$. The ion at m/e 127 is a loss of 15 mass units from the molecular ion and indicates the presence of a methyl group. The small ion at m/e 112 was found to be $C_6H_8O_2$ corresponding to the loss of OCH_2 from the molecular ion. The ions at m/e 99, 71, and 69 were found by mass measurement to be losses of CH_3CO, $C_3H_3O_2$, and $C_3H_5O_2$, respectively, from the molecular ion. These data suggested a homolog of the furanone compound discussed in the previous paragraph. The molecular weight of this compound is 14 mass units higher than 2,5-dimethyl-4-hydroxy-3(2H)-furanone, which points to the presence of an additional methyl group in place of a hydrogen atom. Once again other spectral data are required to define the structural configuration of this molecule. NMR data indicated the presence of a methyl group, a methyoxyl group, and a $CHCH_3$ group. IR data indicated that a conjugated carbonyl group was present in the molecule. From these combined data it was postulated that this strawberry flavor component was 2,5-dimethyl-4-methoxy-3(2H)-furanone [28]. This compound was synthesized and the identification was confirmed since its mass spectrum, NMR spectrum, IR spectrum, GC retention time, and organoleptic properties were identical to the strawberry flavor component.

Pyrazines represent a class of compounds which make very important and distinctive contributions to food flavors. They are usually described as possessing strong nutty aromas but many variations of aroma occur within this class of compounds. An excellent review article on pyrazines in foods lists the pyrazine compounds which have been identified in foods and discusses their organoleptic properties, their isolation procedures, their synthesis techniques, and their spectral data [29]. An extremely potent chemical, 2-methoxy-3-isobutylpyrazine, was reported as the compound which gives bell peppers their unique flavor [30].

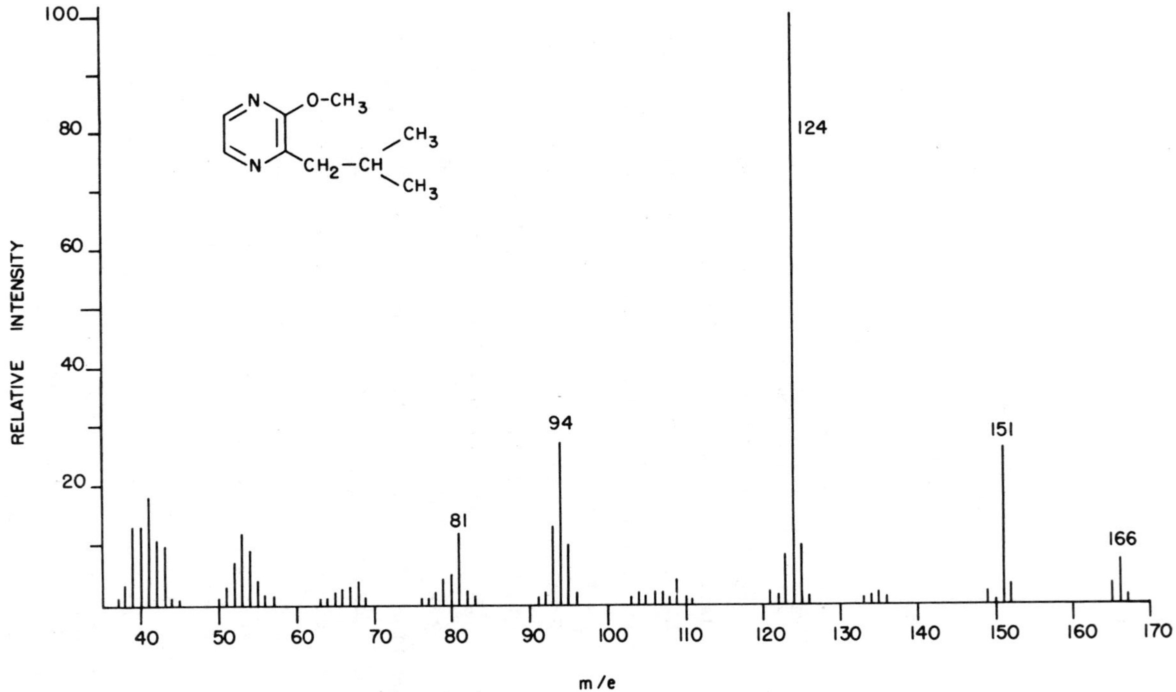

FIG. 16. Mass spectrum of 2-methoxy-3-isobutylpyrazine: DuPont 21-491, 70 eV, GC-MS inlet system.

This identification was corroborated by independent work done in the author's laboratory during the same time period.

The mass spectrum of this compound is shown in Figure 16. The molecular ion at m/e 166 was shown by high-resolution mass spectral analysis to have an empirical formula of $C_9H_{14}N_2O$. An alkyl pyrazine was suspected because of the prominent ion at m/e 94 which is a characteristic ion in the spectra of alkyl pyrazines containing three or more carbons. It is formed by means of a McLafferty rearrangement process (1). The ion at m/e 135

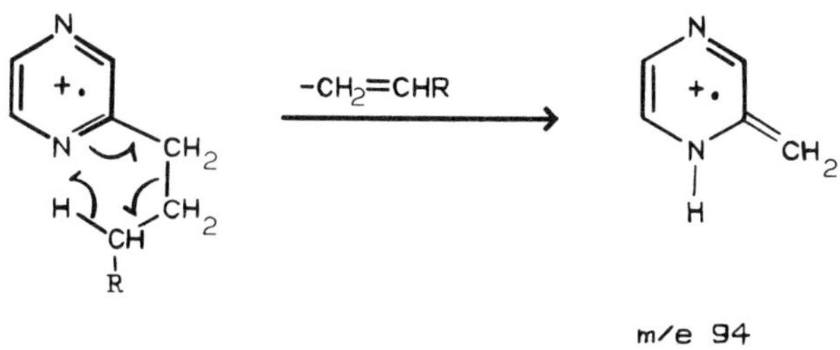

(1)

was found to have a formula of $C_8H_{11}N_2$ and thus was formed by fragmentation of a methoxyl group from the molecular ion. The base peak in the spectrum at m/e 124 has an empirical formula of $C_6H_8N_2O$ and a metastable ion (m*) at 92.6 indicated that it was formed directly from the molecular ion. From these data it was concluded that the unknown compound was a pyrazine substituted by a methoxyl group and a four-carbon alkyl group side chain, and that the base peak at m/e 124 was produced by a McLafferty rearrangement process (2).

The position of the methoxy group with respect to the four-carbon alkyl group could not be predicted from the mass spectrum. The structure of the alkyl substituent was not known for certain although an isobutyl group was suspected because of the prominent ion at m/e 151 which corresponds to the loss of a methyl

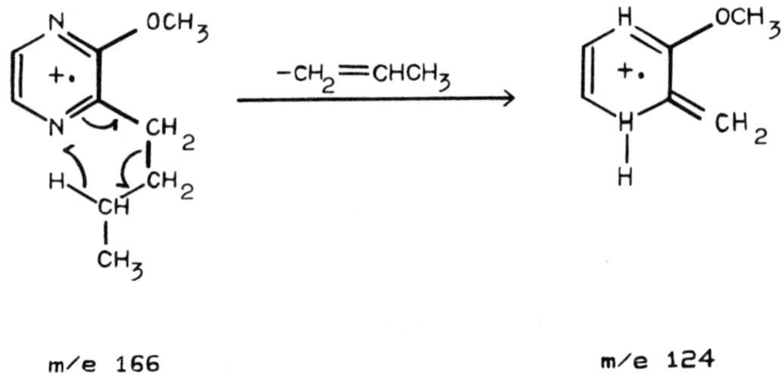

(2)

group from the molecular ion. This process occurs much more abundantly from an isobutyl group than from an n-butyl group. NMR data were used to confirm the presence of an isobutyl group on the ring carbon adjacent to the ring carbon containing the methoxy group.

An example of methoxy-alkylpyrazine isomers where the 2, 3 isomer and the 2,6 isomer can be readily distinguished from their mass spectra is demonstrated by the methoxy-methylpyrazine isomers [31]. The mass spectrum of the 2,3 isomer shows the expulsion of the molecule of water from the molecular ion by means of the "ortho" effect. This process is not observed in the spectrum of the 2,6 isomer. This same effect is observed in the corresponding methylthio-methylpyrazines, where the expulsion of the molecule of H_2S from the molecular ion occurs in the mass spectrum of the 2,3 isomer but not in the spectrum of the 2,6 isomer [31].

Isoamyl acetate has a fruity, pearlike aroma and when diluted it is reminiscent of pears, overripe bananas, and apples. It is widely distributed in nature and has been identified in 12 fruits, 9 alcoholic beverages, cheese, cocoa, coffee, tea, peanuts, and honey. As already mentioned, esters are a class of compounds very prevalent in foods and frequently make important contributions to their flavor. Correlation studies of

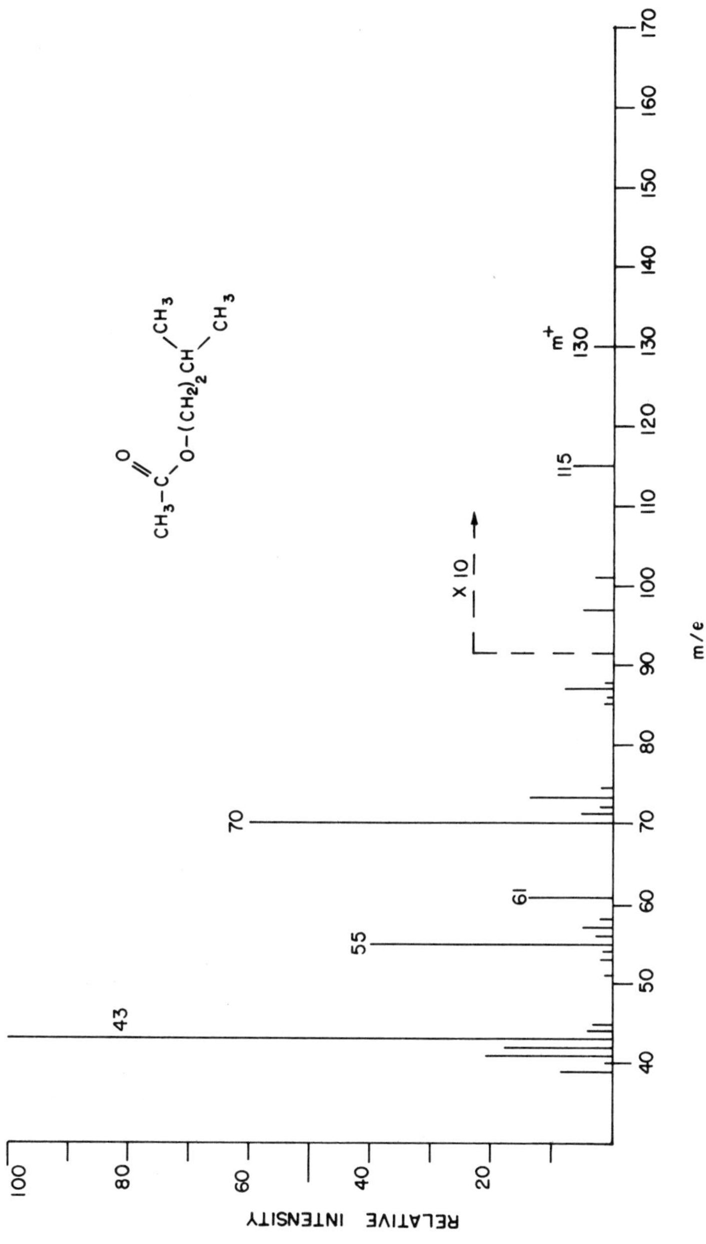

FIG. 17. Mass spectrum of isoamyl acetate: DuPont 21-491, 70 eV, GC-MS inlet system.

mass spectra with molecular structure for the esters of aliphatic acids have been published [32,33]. Esters do not have large molecular ions in their mass spectra. In fact, many times little or no molecular ion is observed. The molecular ion in the mass spectrum of isoamyl acetate (Fig. 17) at m/e 130 is only 0.3% relative intensity and may go undetected where the compound is present in a flavor isolate at low concentration and a weak spectrum is obtained. Nevertheless, the prominent ion at m/e 61 is characteristic of acetates and strongly suggests the presence of an ester. This ion results from cleavage of the alkyl oxygen bond with rearrangement of two hydrogen atoms to the acidic portion of the molecule which retains the positive charge. The intense ion at m/e 70 indicates that the alcohol portion of the molecule contains a five-carbon moiety because this is a characteristic rearrangement ion in the mass spectra of pentyl esters. Fragmentation of the alkyl oxygen bond is also responsible for the m/e 70 ion; however, this time only one hydrogen migrates to the acidic portion of the molecule and charge retention is retained by the alkyl moiety. These data establish that the mass spectrum is from a five-carbon alkyl acetate. The fact that this molecule is actually the isoamyl isomer cannot be ascertained with absolute certainty, but other spectral data and GC retention time data can be used to determine the exact structure of the alkyl portion of the molecule.

The internal esters of gamma- and delta-hydroxy acids are frequently found in food products and contribute distinctive flavor characteristics. The mass spectra of lactones have been studied in detail [34,35]. The prevalent fragmentation found in these spectra is cleavage of the carbon-carbon bond alpha to the ring, yielding an intense ion at m/e 85 for gamma-lactones and an intense ion at m/e 99 for delta-lactones (3). The delta-lactone of 5-hydroxydecanoic acid (delta decalactone) possesses a fruity, coconut-like aroma and has been reported

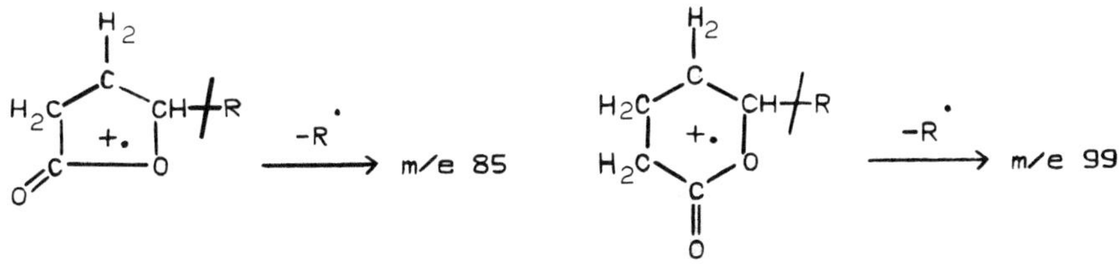

(3)

in apricot, peach, strawberry, cheese, meat, and alcoholic beverages [8]. Figure 18 represents the mass spectrum of delta-decalactone. The dominant ion at m/e 99 strongly suggests the presence of a delta-lactone. Both gamma- and delta-lactones have larger molecular ions than those found in esters; however, the molecular ion is generally present at levels of only a few percent relative intensity. The molecular ion in this spectrum at m/e 170 (rel. int. = 1.3%) and the ion at m/e 152 (rel. int. = 4%), which is formed by the characteristic loss of water from the molecular ion of lactones, show that this compound is actually delta-decalactone. Sometimes a very small amount of compound is isolated from a flavor sample and a weak spectrum is obtained without a discernable molecular ion, but the presence of a lactone is indicated by an intense ion at m/e 85 or m/e 99. In these cases gas chromatographic retention time data can be used to distinguish between the lactones.

Furan compounds are widely distributed in food flavor systems, especially in heat-treated foods where compounds such as furfuryl alcohol, furfural, and hydroxy methyl furfural are produced by nonenzymatic browning reactions. Furfuryl mercaptan has been identified as a component of coffee flavor [36]. It has a penetrating, sulfury, unpleasant odor, but at concentrations below 5 ppb it is described as having a caramel, coffee-like character. The mass spectrum of furfuryl mercaptan (Fig. 19) is relatively simple with major ions at m/e 114, 81, and 53.

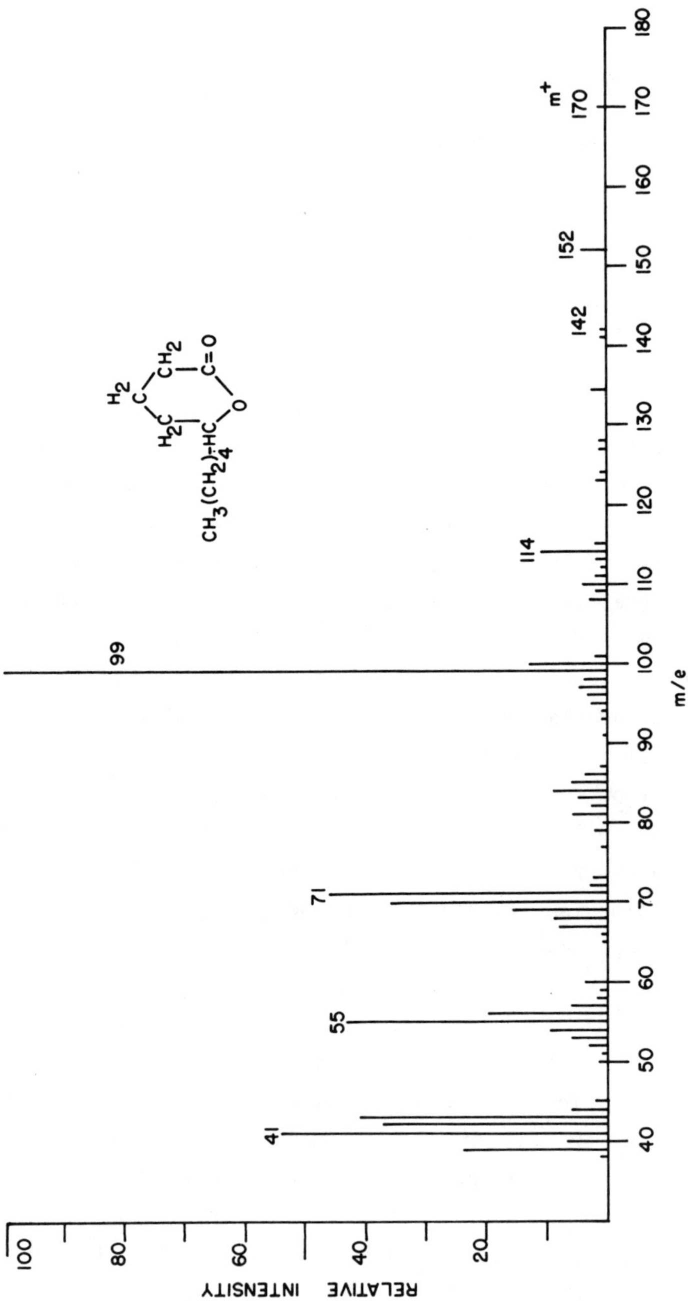

FIG. 18. Mass spectrum of delta-decalactone: DuPont 21-491, 70 eV, GC-MS inlet system.

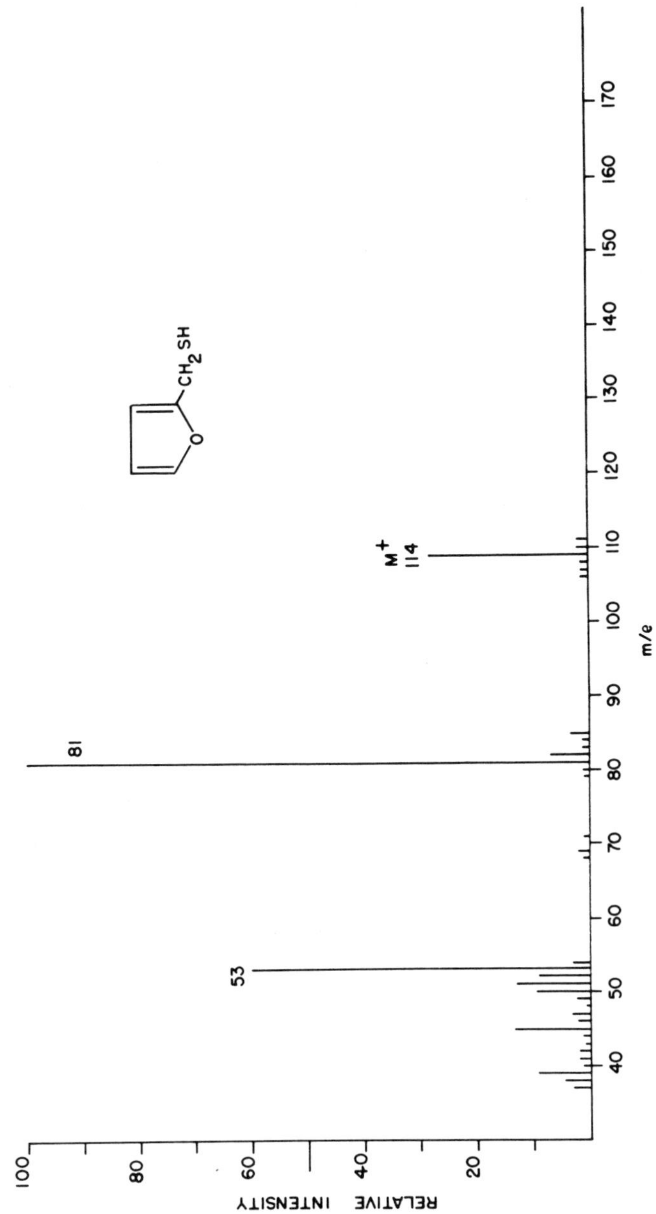

FIG. 19. Mass spectrum of furfuryl mercaptan: DuPont 21-491, 70 eV, GC-MS inlet system.

The molecular ion at m/e 114 is immediately suspected of containing a sulfur atom because the ion at m/e 116, which is two mass units higher than the molecular ion (M + 2), is approximately 4% to 5% of the intensity of the molecular ion. Sulfur is easily recognized in the mass spectrum of an organic molecule because of the natural abundance of the S^{34} isotope (4.2%) which gives a peak two mass units higher than the molecular ion. The presence of sulfur in the molecule is supported by the loss of 33 mass units directly from the molecular ion to give a peak at m/e 81. Loss of 33 mass units directly from the molecular ion is verified by a metastable ion (m*) at 57.6. This is a characteristic fragmentation pattern of thiols where an SH group is lost from the molecular ion. The base peak in the spectrum at m/e 81 is a characteristic fragment ion in furfuryl compounds. It is formed by cleavage of the carbon bond beta to the furan ring and establishes that a methylene group is joining an SH group to a furan ring. The presence of the furan ring is supported by an intense ion at m/e 53 ($C_4H_5^+$).

Allyl isothiocyanate has a sharp, pungent, horseradish, mustard-like aroma and is found in cabbage, onion, garlic, and mustard [8]. Its mass spectrum is shown in Figure 20. Once again the presence of sulfur in the molecule is indicated by the abundance of the M + 2 ion at m/e 101. The odd-numbered molecular ion at m/e 99 indicates the presence of a nitrogen atom in the molecule. The prominent ion at m/e 72 is a characteristic fragmentation ion in the spectra of isothiocyanates. It is formed by cleavage of the bond beta to the functional group and is composed of $(CH_2\text{-}N{=}C{=}S)^+$. The isothiocyanate structure is strongly indicated since an ion at m/e 72 is not characteristic of other sulfur containing compounds. The base peak in the spectrum at m/e 41 ($C_3H_5^+$) indicates unsaturation. Subtraction of the ion at m/e 72 from the molecular ion at m/e 99 leaves 27 mass units to be accounted for, which correspond

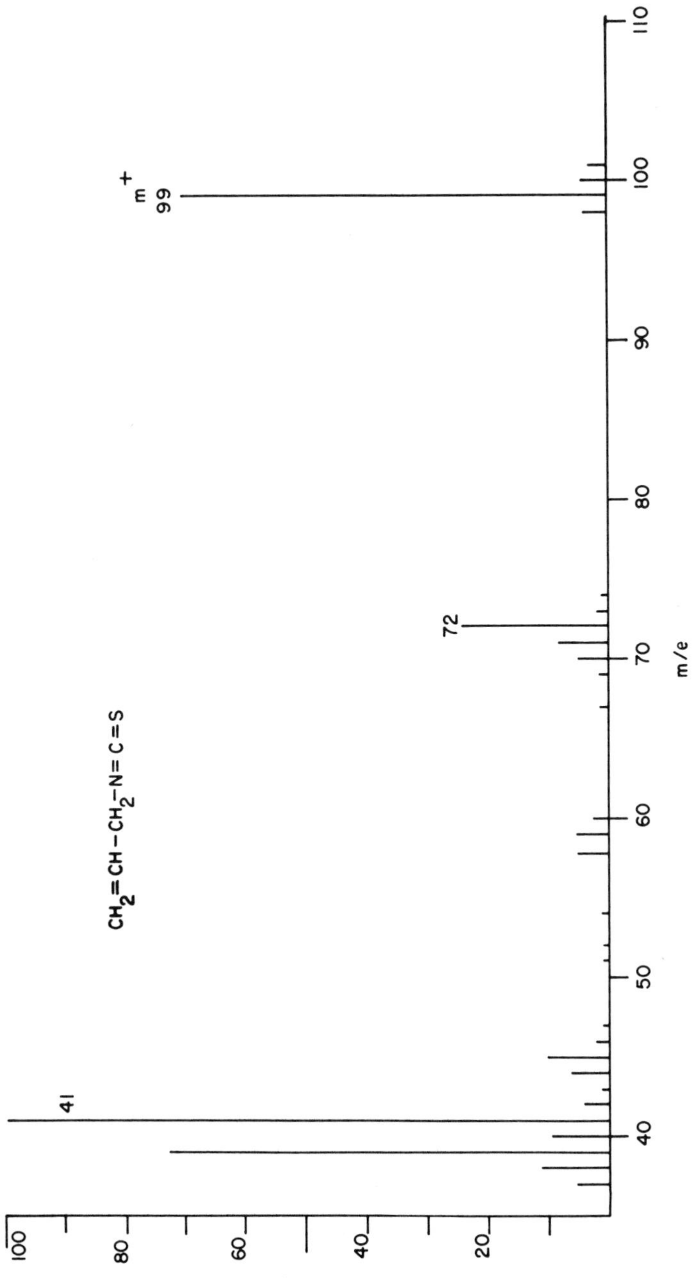

FIG. 20. Mass spectrum of allyl isothiocyanate: DuPont 21-491, 70 eV, GC-MS inlet system.

to a CH_2=CH- group. Joining this group to the methylene group (m/e 72) leads to the conclusion that the spectrum is allyl isothiocyanate.

The preceding examples are not meant to be a comprehensive discussion of the types of structures encountered in flavor isolates, but rather are intended to give the reader a feel for the techniques and thought processes used in the interpretation of mass spectra. In many cases molecular weight data, correlations of structure with fragment ions including specific rearrangement processes, isotopic abundance data, and the nitrogen rule can be used to elicit a good description of the molecule, if not the exact structure. Obviously, comparison of unknown mass spectra to a standard library of spectra (whether by hand or computer) and high-resolution data add new dimensions to the interpretation of the spectrum, and greatly enhance the probability of complete structure elucidation. In those cases where the complete structure cannot be defined solely from the mass spectrum, use of NMR, IR, and UV spectral data along with gas chromatographic retention time data will usually supply the missing pieces of information.

IV. MASS SPECTRA OF NONVOLATILE FOOD COMPOUNDS

Besides the flavor compounds occurring naturally in foodstuffs, a wide variety of flavors are produced during the processing and heat treatment of food products. The reactions which are responsible for the production of these flavors are classified as nonenzymatic browning reactions and are accompanied by discoloration or browning of the food. These flavor changes are usually beneficial to the flavor of a food as in the mouthwatering aromas emanating from a piece of beef being roasted or a loaf of bread being baked. However, in some cases nonenzymatic browning reactions are responsible for off-flavors in a food product.

The chemistry of nonenzymatic browning reactions in the production of flavor compounds in food products [37,38] is a very broad field of study and encompasses a variety of chemical reactions. Caramelization of sugars and other polyhydroxy-carbonyl compounds occurs at elevated temperatures with the production of many flavorful compounds including aldehydes, ketones, acids, furans, and alcohols. The Maillard reaction involves the reaction of carbonyl compounds such as reducing sugars with the free amino group of amines, amino acids, peptides, and proteins through a complex series of chemical pathways to produce a wide variety of flavor compounds including: carbonyls, sulfides, thiols, and volatile nitrogen compounds. The Strecker degradation is one major pathway of the Maillard reaction which results in the heat-induced degradation of alpha-amino acids in the presence of dicarbonyl compounds to produce volatile aldehydes containing one less carbon atom. Oxidation reactions involving lipid materials are another main source of flavor components in foods and result in the production of such volatile compounds as acids, carbonyls, lactones, and alcohols.

The importance of the nonvolatile, high-molecular-weight compounds such as sugars, amino acids, peptides, proteins, and lipids as precursors of the complex variety of flavor compounds in foodstuffs is well established. Analysis and identification of these compounds, therefore, is an integral part of food and flavor chemistry. Reviews have appeared on the mass spectral analysis of carbohydrates [39], amino acids [40], amino acid sequence in oligopeptides [41], fatty acids [42], and complex lipids [43]. A variety of techniques including esterification, methylation, and trimethylsilylation have been developed to render these compounds more volatile and thus amenable to isolation from a complex food system by gas chromatography and identification by mass spectrometry. Since these derivatives have a large molecular weights and the mass spectra are quite

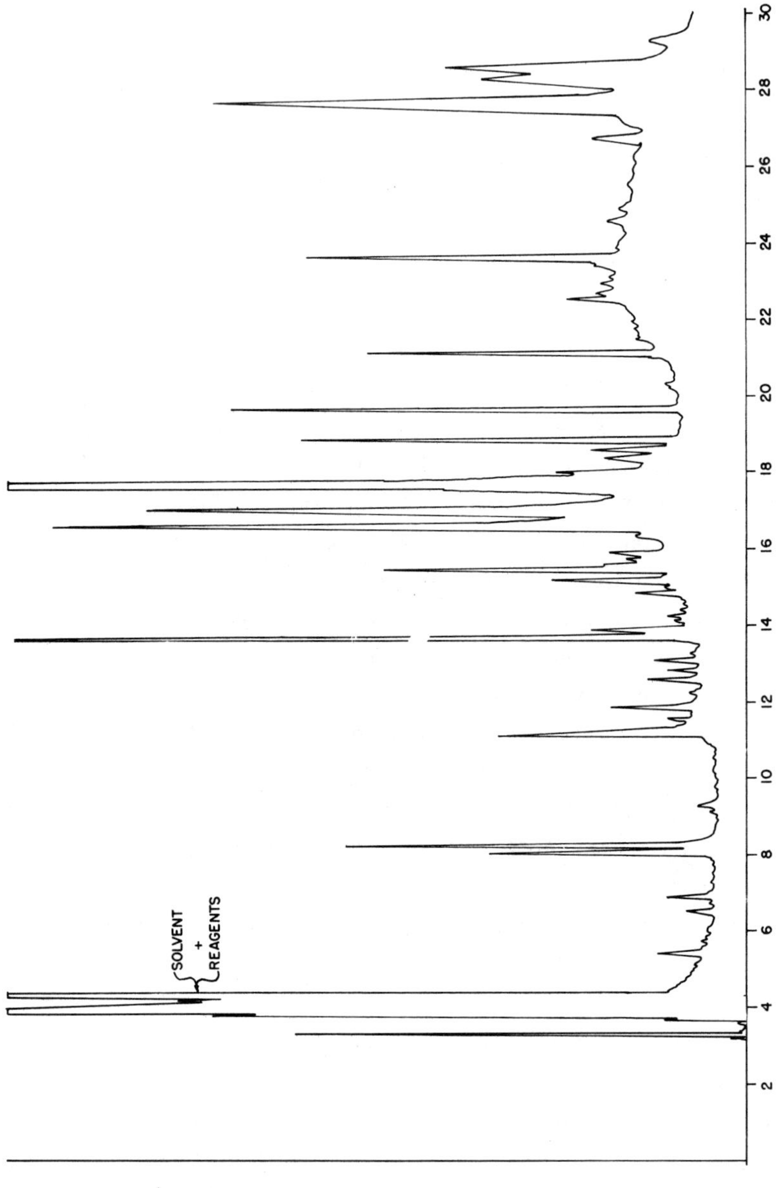

FIG. 21. Gas chromatogram of a coffee nonvolatiles fraction (Trimethylsilylethers) from a 500 ft x 0.03 in. id capillary column treated with OV-101. Programmed from 70°C to 300°C at 12°/min.

complex, it is desirable and often essential to obtain high-resolution mass spectral data for structure elucidation.

A gas chromatogram of the trimethylsilyl (TMS) derivatives of the nonvolatile components isolated from coffee is shown in Figure 21. Formation of the TMS derivatives of the nonvolatile compounds containing hydroxyl, carboxyl, and amino groups has the advantage that the derivatives are easily prepared and the volatility of the TMS compounds is compatible with GC-MS analysis. A component of the coffee isolate was identified as the TMS derivative of α-D-glucose. Its GC retention time was 18.3 min and its mass spectrum is shown in Figure 22. The prominent ions at m/e 73 and m/e 147 are characteristic ions in the spectra of TMS derivatives. The ion at m/e 73 represents simple

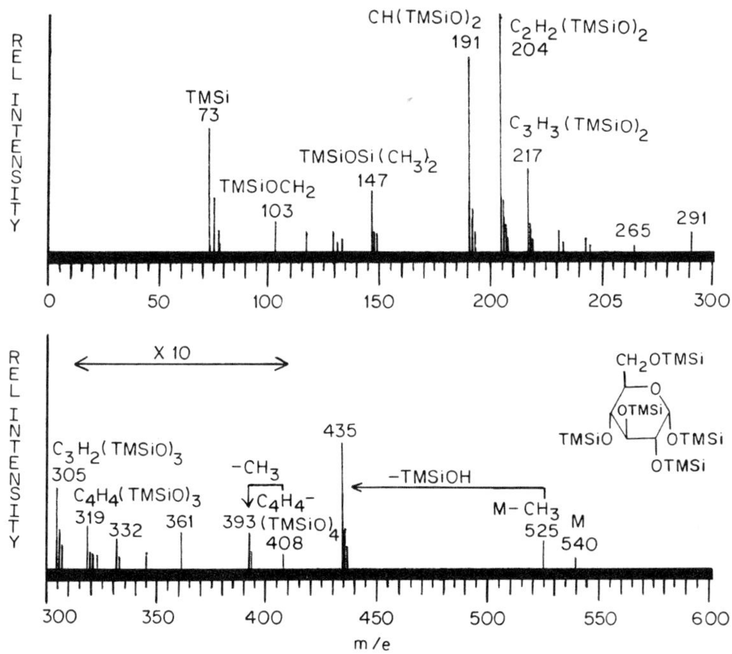

FIG. 22. Mass spectrum of α-D-glucose (TMS): CEC 21-110, 70 eV, direct probe inlet.

cleavage of a TMS group $[Si(CH_3)_3^+]$. The ion at m/e 147 is the result of a complex rearrangement process involving two TMS groups which form a $[CH_3)_2SiOSi(CH_3)_3^+]$ ion. Analysis of this spectrum shows that a small molecular ion is present at m/e 540. A very characteristic fragmentation observed in the mass spectra of TMS compounds is the loss of a methyl group from the molecular ion at m/e 525 $(M-CH_3)$. The ion at $(M - CH_3)$ is often larger than the molecular ion and is used to determine the molecular weight of those TMS compounds which exhibit little or no molecular ion. A molecular weight of 540 combined with intense ions at m/e 204, 191, and 217 indicate that the mass spectrum represents a TMS hexose. The base peak in this spectrum is at m/e 204 and is produced by ring cleavage. It contains two ring carbons and two trimethylsiloxy (TMSO) groups. A base peak at m/e 204 indicates the pyranose form of the hexose while a base peak at m/e 217 would indicate the furanose form. The prominent ion at m/e 217 is composed of three ring carbons and two TMSO groups. The intense rearrangement ion at m/e 191 also involves ring cleavage and is composed of one ring carbon and two TMSO groups.

The prominent rearrangement ion at m/e 435 is characteristic of the mass spectra of TMS aldohexoses. It is formed by expulsion of a molecule of trimethylsilanol from the $(M - CH_3)$ ion. This important fragment distinguishes the spectra of aldohexoses from the spectra of ketohexoses which do not have an ion at m/e 435. Ketohexoses tend to eliminate a $CH_2OSi(CH_3)_3$ group directly from the molecular ion producing a prominent ion at m/e 437. Another prominent ion in this mass spectrum is observed at m/e 408 and is composed of four ring carbons with their respective TMSO groups. Loss of a methyl group from this ion produces the fragment at m/e 393. The ions at m/e 305 and 319 represent three ring carbons with three TMSO groups and four ring carbons with three TMSO groups, respectively. Extensive rearrangement processes are observed in the spectra of TMS derivatives. This is illustrated by the small intensity ion at m/e 265 which is composed of two intact TMSO groups and one TMSO group minus a methyl group, attached to a single ring carbon. Analysis of this spectrum using known correlation ions establishes that it is the spectrum of the pyranose form of an aldohexose. Since the mass spectrum does not distinguish the epimeric or anomeric forms, gas chromatography retention time data were used to establish that this component was α-D-glucose.

Studies of the structure of correlation ions in the TMS mass spectra of carbohydrates, which can be used to identify and distinguish between classes of carbohydrates, have been published [39,44-48]. As already mentioned aldohexoses can be differentiated from ketohexoses by the ion at m/e 435 for aldohexoses and m/e 437 for ketohexoses. Five- and seven-carbon sugars have TMS mass spectra similar to the hexoses but can be differentiated by the difference in their molecular ions and ($M-CH_3$) ions. Alditols, such as glucitol and mannitol, have distinguishing ions at m/e 319, 307, and 205 while the fragments at m/e 204 and 191 found at high intensity in hexoses are at much lower intensity in the spectra of alditols. Methylglycosides can be distinguished from hexoses by their molecular ions at m/e 482. Similar to the spectra of hexoses, the methyl glycosides of aldohexoses lose a trimethylsilanol molecule from the ($M - CH_3$) ion forming a prominent m/e 377, while methyl glycosides of ketohexoses lose a $CH_2OSi(CH_3)_3$ moiety directly from the molecular ion producing an abundant ion at m/e 379. Glucosamine, an example of 2-amino-2-deoxyhexoses, where the four hydroxyl groups are trimethylsilylated, is characterized by an intense ion at m/e 131. Glucuronic acid and other uronic acids of aldohexoses can be recognized by intense ions in the m/e range from 215 to 235. These examples represent a cross section of the correlation ions found in the mass spectra of carbohydrate

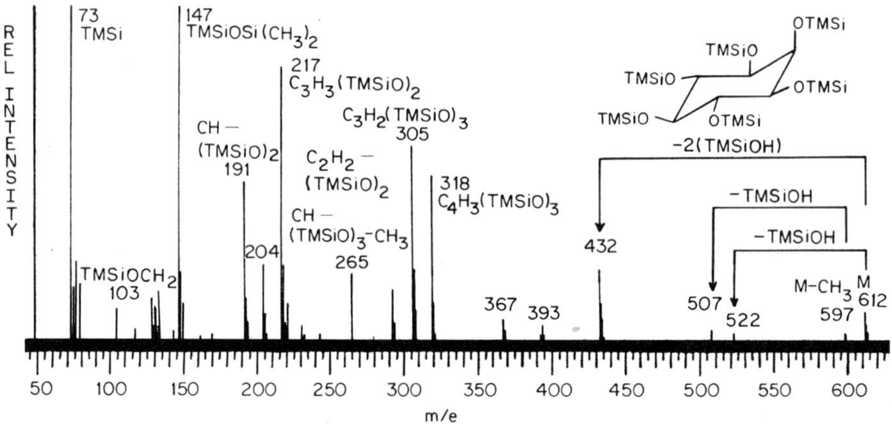

FIG. 23. Mass spectrum of myo-inositol (TMS): CEC-21-110, 70 eV, direct probe inlet.

TMS derivatives which permit classification of the carbohydrate type. When these data are complemented by GC retention-time data, it will usually define their isomeric configuration.

Myo-inositol is widely distributed in nature having been identified in various plants and animals. It has been reported as a component of roasted coffee [49]. The mass spectra of the TMS derivatives of the eight inositol diastereoisomers have been studied in detail [50]. Large variations in the ion intensities between the mass spectra of the different inositol isomers were observed. The component having a retention time of 19.7 min in the gas chromatogram of the coffee nonvolatiles (Fig. 21) was identified as myo-inositol. Its mass spectrum is shown in Figure 23. A prominent molecular ion is observed at m/e 612 and a less intense fragment is found at m/e 597 ($M-CH_3$). The characteristic ions of TMS derivatives are found at m/e 73 and m/e 147. A rearrangement process involving the loss of a molecule of trimethylsilanol (TMSOH) from the molecular ion is characteristic of the spectra of many TMS compounds. Losses of TMSOH from both the molecular ion and the ($M-CH_3$) ion

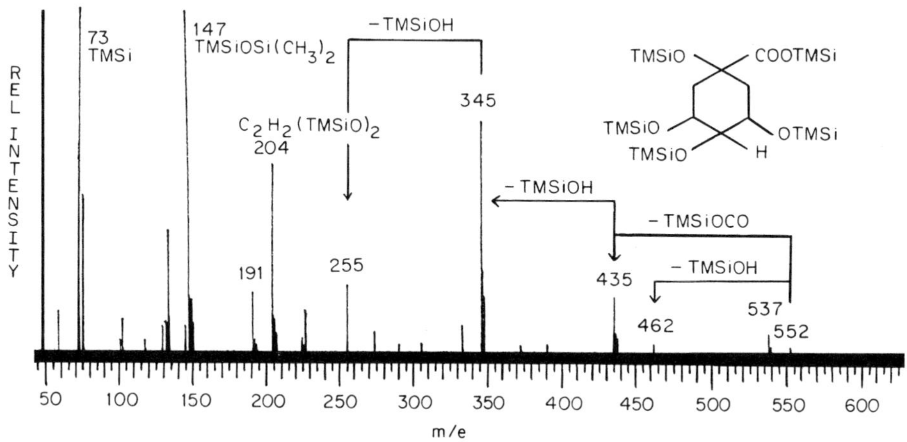

FIG. 24. Mass spectrum of quinic acid (TMS): CEC 21-110, 70 eV, direct probe inlet.

are noted in this spectrum at m/e 522 and at m/e 507. The abundant ion at m/e 432 is formed by the expulsion of two TMSOH molecules from the molecular ion. Intense fragments occur at m/e 305 (three ring carbons and three TMSO groups) and at m/e 318 (four ring carbons and three TMSO groups). The extensively rearranged ion at m/e 265 is more intense in this spectrum than in the mass spectrum of α-D-glucose.

Quinic acid is commonly found in nature and was identified as a component of the coffee nonvolatiles, having a retention time of 17.7 min (Fig. 21). The mass spectrum of the TMS derivative of quinic acid is shown in Figure 24. A very small intensity molecular ion is observed at m/e 552, with a more abundant ion being found at m/e 537 (M - CH_3). An interesting fragmentation series is demonstrated by this mass spectrum. First, the trimethylsilylated carboxylic acid (TMSOCO) moiety is lost from the molecular ion (M -TMSOCO) producing the prominent ion at m/e 435. This ion then loses a molecule of trimethylsilanol forming a very intense m/e 345. This ion in turn loses another molecule of TMSOH producing the ion at m/e 255. A

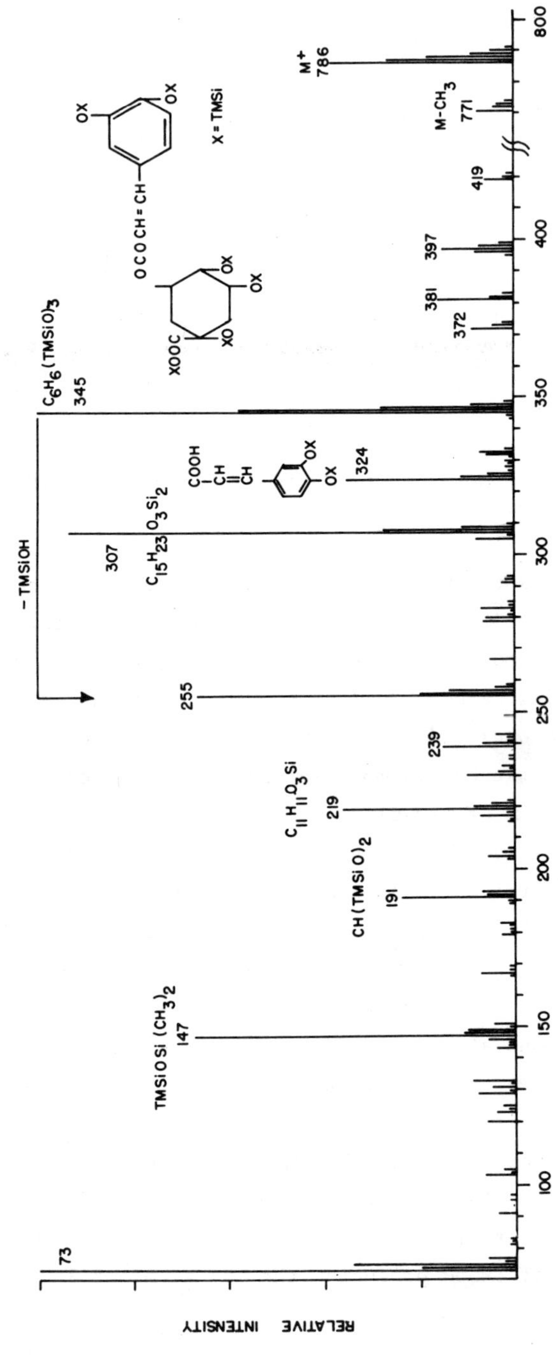

FIG. 25. Mass spectrum of chlorogenic acid (TMS): CEC 21-110, 70 eV, direct probe inlet.

molecule of TMSOH is also expelled directly from the molecular ion to give a small m/e 462. The ion at m/e 191 is prominent in this spectrum just as it is in the spectra of α-D-glucose and myo-inositol; however, the ions at m/e 204 and 217 are very small in this spectrum.

Chlorogenic acid is phenolic acid which is widely distributed in nature. Known as caffeoylquinic acid, it is the ester of a molecule of caffeic acid and a molecule of quinic acid. It was identified in the coffee nonvolatiles (Fig. 21) as the component at retention time 27.6 min. The mass spectrum of the TMS derivative of chlorogenic acid is given in Figure 25 [51]. An intense molecular ion is found at m/e 786 in the spectrum of this molecule, which contains six TMS groups. The loss of a methyl group from the molecular ion shows up as m/e 771. The intense m/e 345 is the same ion as was observed in the spectrum of quinic acid and obviously is produced from the quinic acid portion of the molecule. As in the case of the quinic acid spectrum, a molecule of TMSOH is lost from m/e 345 forming the ion at m/e 255. The rearrangement ion at m/e 324 represents the caffeic acid portion of the molecule and is composed of a molecule of caffeic acid with both of the hydroxyl groups retaining their TMS substituents. High-resolution data indicate that the intense ion at m/e 307 is produced by cleavage of the carbon-oxygen bond beta to the quinic acid ring with charge retention on the caffeic acid portion of the molecule. The prominent m/e 219 also appears to be from the caffeic acid portion of the molecule. It is probably formed by cleavage of the carbon-oxygen bond beta to the quinic acid ring accompanied by loss of a TMS group and loss of a single methyl group from the other OTMS group on the caffeic acid moiety.

V. FUTURE OF MASS SPECTROMETRY IN THE FOOD INDUSTRY

Mass spectrometry has carved a permanent niche in the food and flavor laboratory and in light of recent trends should become an even more potent part of food analysis. As in many other fields of research, computers have proven themselves to be extremely powerful tools in the mass spectrometry laboratory dedicated to food research. First, a great deal of time is saved in the storing and processing of the volumes of data retrieved from GC-MS systems. Also, with the wide variety of search routines available, a large amount of data inherently present in the mass spectrum, which were extremely difficult if not impossible to dig out by manual techniques, surface and become usable. Computer technology was first applied to mass spectrometry in the processing of high-resolution data [52]. Soon after, its potential for acquisition and reduction of the vast amount of data produced by combined gas chromatography-mass spectrometry systems was recognized [53].

Comprehensive reviews of computer acquisition, processing, and analysis of mass spectral data [54,55] and the role of the computer in GC-MS systems [56] have been written. A combined GC-MS-computer system dedicated to the analysis of volatile food components was described at a recent symposium on computers in flavor chemistry [57]. Besides the acquisition, reduction, and storage of mass spectral data, computer files of standard mass spectra are commercially available (to which one's own data can be added) for searching, matching, and identifying mass spectra. Methods are presently being studied for the correlative and deductive interpretation of mass spectra by computer; however, this process is not clearly defined since years of experience by the skilled mass spectrometrist go into the interpretation of mass spectral data, which in many respects is an acquired art. In the final analysis, the alliance between gas chromatography and mass spectrometry becomes an even more formidable analytical tool with the inclusion of data handling systems.

Electron impact along with chemical and field ionization are well-established techniques for the production of ions in a mass spectrometer. A newer technique with exciting horizons for the analysis of food components having large organic molecules with low vapor pressures is field desorption mass spectrometry (see Chapter 5) [58,59]. This technique involves the deposition of the solid organic molecules on the field anode which is inserted into the ionization chamber of the mass spectrometer. Ionization of the sample takes place by desorption directly from the anode at high electric field strength without the need for high-temperature vaporization, where thermal degradation is always a risk. This facilitates the analysis of underivatized amino acids, peptides, saccharides, and other nonvolatile compounds found in food systems. Although still in its developmental stages and not available in the average food laboratory at the present time, this technique holds a great deal of promise for future food and flavor research.

Atmospheric pressure ionization [60] is another new technique with promising potential in the area of biochemical and food research. Ionization of the sample molecules takes place in a heated source external to the mass spectrometer at atmospheric pressure. Ions are formed by a complex series of ion-molecule reactions in a flowing gas or liquid stream and enter the mass analyzer through a small aperture. This technique appears to be well suited for the interfacing of high-pressure, high-resolution liquid chromatography systems to a mass spectrometer [61]. In this way, the eluant stream from a liquid chromatograph is continuously analyzed by the mass spectrometer. Just as tandem GC-MS analysis has contributed so much information about volatile flavor components, liquid chromatography combined with mass spectrometry should make a significant contribution to the analysis of the less volatile components found in food systems.

Mass spectrometry has made, is making, and will continue to make very significant contributions in the area of food and flavor research. The presence of a mass spectrometer in today's laboratory dedicated to food and flavor research is commonplace. The relatively high initial cost of a mass spectrometer is amply offset by the vast amount of useful data it produces in relatively short periods of time, and by the fact that in many instances it may be the only route to the successful completion of a research project. It is safe to assume that the future thrust in food research to more completely understand the very complex biochemical processes taking place in food systems will depend heavily on mass spectrometry as an essential analytical tool.

REFERENCES

1. C. Kleber, *Amer. Pefum.*, *7*, 235 (1913).
2. R. G. Buttery, R. M. Seiffert, D. G. Guadagni, and L. C. Ling, *J. Agr. Food Chem.*, *17*, 1322 (1969).
3. W. H. McFadden, *Techniques of Combined Gas Chromatography/Mass Spectrometry: Applications in Organic Analysis*, Wiley-Interscience, New York, 1972.
4. W. H. Stahl (Ed.), *Compilation of Odor and Taste Threshold Data*, American Society for Testing and Materials, Philadelphia, 1973.
5. A. Zlatkis, H. A. Lichtenstein, A. Tishbee, W. Bertsch, and H. M. Liebich, *J. Chromatogr. Sci.*, *11*, 299 (1973).
6. C. G. Tassan and G. F. Russell, *J. Food Sci.*, *39*, 64 (1974).
7. T. H. Parliment, W. Clinton, and R. Scarpellino, *J. Agr. Food Chem.*, *21*, 485 (1973).
8. S. Van Straten and F. de Vriger, *List of Volatile Compounds in Foods, 3rd Ed., Report No. R4030*, Central Institute for Nutrition and Food Research, Utrecht, The Netherlands, 1973.
9. K. Biemann, *Mass Spectrometry: Organic Chemical Applications*, McGraw-Hill, New York, 1962.
10. F. W. McLafferty (Ed.), *Mass Spectrometry of Organic Ions*, Academic Press, New York, 1963.
11. F. W. McLafferty, *Mass Spectral Correlations*, Advances in Chemistry Series 40, American Chemical Society, Washington, 1963.

12. H. Budzikiewicz, C. Djerassi, and D. H. Williams, *Mass Spectrometry of Organic Compounds*, Holden-Day, San Francisco, 1967.

13. J. H. Beynon, R. A. Saunders, and A. E. Williams, *The Mass Spectra of Organic Molecules*, Elsevier, Amsterdam, 1968.

14. G. R. Waller (Ed.), *Biochemical Applications of Mass Spectrometry*, Wiley-Interscience, New York, 1972.

15. M. C. Hamming and N. G. Foster, *Interpretation of Mass Spectra of Organic Compounds*, Academic Press, New York, 1972.

16. Q. N. Porter and J. Baldas, *Mass Spectrometry of Heterocyclic Compounds*, Wiley-Interscience, New York, 1971.

17. M. G. Kolor, in *Biochemical Applications of Mass Spectrometry*, (G. R. Waller, Ed.), Wiley-Interscience, New York, 1972, Chap. 25.

18. W. H. McFadden, in *Topics in Organic Mass Spectrometry* (A. L. Burlingame, Ed.), Wiley-Interscience, New York, 1970, Chap. 8.

19. R. Teranishi, I. Hornstein, P. Issenberg, and E. L. Wick, *Flavor Research: Principles and Techniques*, Dekker, New York, 1971, Chaps. 5 and 6.

20. M. G. Kolor, in *Biochemical Applications of Mass Spectrometry* (G. R. Waller, Ed.), Wiley-Interscience, New York, 1972, pp. 712-714.

21. R. W. Holley, B. Stoyla, and A. D. Holley, *Food Res., 20*, 326 (1955).

22. H. Schniz and C. F. Seidel, *Helv. Chim. Acta, 44*, 278 (1961).

23. M. Winter and E. Sundt, *Helv. Chim. Acta, 45*, 2195 (1962).

24. K. Biemann, *Mass Spectrometry: Organic Chemical Applications*, McGraw-Hill, New York, 1962, pp. 103-105.

25. A. F. Thomas, B. Willhalm, and R. Muller, *Org. Mass Spectrom., 2*, 223 (1969).

26. J. O. Rodin, C. M. Himel, R. M. Silverstein, R. W. Leeper, and W. A. Gortner, *J. Food Sci., 30*, 280 (1965).

27. L. Re, B. Maurer, and G. Ohloff, *Helv. Chim. Acta, 56*, 1882 (1973).

28. M. G. Kolor, unpublished work, 1974.

29. J. A. Maga and C. E. Sizer, *Crit. Rev. Food Technol., 4*, 117 (1973).

30. R. G. Buttery, R. M. Seifert, R. E. Lundin, D. G. Guadagni, and L. C. Ling, *Chem. Ind. (London)*, *17*, 490 (1969).

31. M. G. Kolor and D. J. Rizzo, *Org. Mass Spectrom.*, *5*, 959 (1971).

32. A. G. Sharkey, J. L. Schultz, and R. A. Friedel, *Anal. Chem.*, *31*, 87 (1959).

33. J. H. Beynon, R. A. Saunders, and A. E. Williams, *Anal. Chem.*, *33*, 221 (1961).

34. W. H. McFadden, E. A. Day, and M. J. Diamond, *Anal. Chem.*, *37*, 89 (1965).

35. E. Honkanen, T. Moisio, and P. Karvonen, *Acta Chem. Scand.*, *19*, 370 (1965).

36. J. Heins, H. Marse, M. C. ten Noever de Brauw, and C. Weurman, *J. Gas Chromatogr.*, *4*, 395 (1966).

37. J. E. Hodge, in *Chemistry and Physiology of Flavors*, (H. W. Schultz, E. A. Day, and L. M. Libbey, Eds.), Avi, Westport, Conn., 1967, Chap. 22.

38. T. M. Reynolds, *Food Tech. Aust.*, *22*, 610 (1970).

39. T. Radford and D. C. DeJongh, in *Biochemical Applications of Mass Spectrometry* (G. R. Waller, Ed.), Wiley-Interscience, New York, 1972, Chap. 12.

40. W. Vetter, in *Biochemical Applications of Mass Spectrometry* (G. R. Waller, Ed.), Wiley-Interscience, New York, 1972, Chap. 14.

41. K. Biemann, in *Biochemical Applications of Mass Spectrometry* (G. R. Waller, Ed.), Wiley-Interscience, New York, 1972, Chap. 15.

42. G. Odham and E. Stenhagen, in *Biochemical Applications of Mass Spectrometry* (G. R. Waller, Ed.), Wiley-Interscience, New York, 1972, Chap. 8.

43. G. Odham and E. Stenhagen, in *Biochemical Applications of Mass Spectrometry* (G. R. Waller, Ed.), Wiley-Interscience, New York, 1972, Chap. 9.

44. O. S. Chizhov, N. V. Molodtsov, and N. K. Kotchetkov, *Carbohy. Res.*, *4*, 273 (1967).

45. D. C. DeJongh, T. Radford, J. D. Hribar, S. Hanessian, M. Bieber, G. Dawson, and C. C. Sweeley, *J. Amer. Chem. Soc.*, *91*, 1728 (1969).

46. J. Karkkainen and R. Vihko, *Carbohy. Res.*, *10*, 113 (1969).

47. G. A. Reineccius, T. E. Kavanagh, and P. G. Keeney, *J. Dairy Sci.*, *53*, 1018 (1970).

48. J. P. Kamerling, J. F. G. Vliegenthart, J. Vink, and J. J. De Ridder, *Tetrahedron, 27*, 4275 (1971).

49. A. R. Mishkin, R. S. Bower, and L. E. Anderson, *Carbohy. Res., 13*, 170 (1970).

50. W. R. Sherman, N. C. Eilers, and S. L. Goodwin, *Org. Mass Spectrom., 3*, 829 (1970).

51. M. G. Kolor, unpublished work, 1973.

52. D. Desiderio and K. Biemann, *12th Annual Conference on Mass Spectrometry and Allied Topics*, Montreal, Canada, 1964.

53. C. Merritt, Jr., P. Issenberg, M. L. Bazinet, B. N. Green, T. O. Merrin, and J. G. Murray, *Anal. Chem., 37*, 1037 (1965).

54. Series of articles on data acquisition and processing in *Biochemical Applications of Mass Spectrometry* (G. R. Waller, Ed.), Wiley-Interscience, New York, 1972, Secs. I and II.

55. G. W. A. Milne (Ed.), *Mass Spectrometry: Techniques and Applications*, Wiley-Interscience, New York, 1971.

56. W. H. McFadden, *Techniques of Combined Gas Chromatography/ Mass Spectrometry: Applications in Organic Analysis*, Wiley-Interscience, New York, 1972, Chap. 7.

57. C. Merritt, Jr., D. H. Robertson, J. F. Cavagnaro, R. A. Graham, and T. I. Nichols, *J. Agr. Food Chem., 22*, 750 (1974).

58. H. D. Beckey, *Int. J. Mass Spectrom. Ion Phys., 2*, 500 (1969).

59. H. D. Beckey and H.-R. Schulten, *Practical Mass Spectrometry, Part A* (C. Merritt, Jr. and C. N. McEwen, Eds.), Marcel Dekker, New York, 1979, Chap. 5.

60. E. C. Horning, M. G. Horning, D. I. Carroll, I. Dzidic, and R. N. Stillwell, *Anal. Chem., 45*, 936 (1973).

61. E. C. Horning, D. I. Carroll, I. Dzidic, M. G. Horning, and R. N. Stillwell, *168th National Meeting of American Chemical Society*, Atlantic City, 1974.

Chapter 4

ANALYSIS OF ENCAPSULATED ENVIRONMENTS

F. E. SAALFELD and R. A. SAUNDERS

Physical Chemistry Branch
Chemistry Division
Naval Research Laboratory
Washington, D. C.

I. INTRODUCTION

Most people know what an encapsulated or enclosed atmosphere is, but few people have experienced one. Spacecraft, submarines, and diving habitats are examples of enclosed or encapsulated environments. It is not generally realized that the personal safety of the people living in these environments depends upon the quality of the air contained in these enclosures. However, every submariner, astronaut, aquanaut, or diver is acutely aware of this fact. Thus, all closed, manned environments have some type of atmosphere purification and control system.

For example, air in the submarine circulates through many purification devices. Electrostatic precipitators and filters remove liquid and solid particles generated from cooking and smoking. A catalytic oxidizer converts carbon monoxide (from smoking) and hydrogen (from battery charging) to carbon dioxide and water. A regenerative organic base removes carbon dioxide from the air by absorption. The carbon dioxide is recovered as a compressed gas and discharged outside of the submarine. Finally a bed of activated charcoal adsorbs the organic contaminants. These systems must work properly if the environment is to remain habitable. Atmosphere analyses are necessary to ensure that this is indeed the case.

One of the scientific areas of interest to the Chemistry Division of the Naval Research Laboratory (NRL) is atmosphere analysis. Submarine atmospheres have been the subject of most of our research; although on occasion we have analyzed others. NRL studies have included aviator's and diver's air supplies [1,2], hyperbaric habitats such as *Sea Lab I*, *II*, and *III*, and *TEKTITE* [3-7], small submersibles [8], space simulators [9], manned and unmanned spacecabins [10], as well as submarines [11-16]. When the NRL submarine atmosphere analysis studies began in the mid-1930s, the underwater endurance capability of a submarine was limited by the capacity of its battery. Nuclear power, however, made long-term submergence possible, and this, in turn, made contamination control essential to the well being of the ship's crew.

All nuclear-powered submarines and some spacecraft have in situ atmosphere analyzers. This is the preferred method of analysis, because it provides current information about the environment's quality. Samples taken from an encapsulated environment and returned to the laboratory for analyses provide information only on the history of the atmosphere, not on its status. However, most of the present-day analyzers used on submarines monitor six or eight constituents depending on the analyzer being used. New instrumental developments indicate that an analyzer can be built which will analyze the complex contaminant mixture in an encapsulated environment and also be sufficiently reliable and rugged for operation under demanding conditions.

It is impossible to anticipate all possible contaminants in these encapsulated environments. The contaminant mixture found in a submarine atmosphere is believed to consist of more than 10,000 species [17]. There are many contaminant sources aboard submarines: crew activities such as smoking and cooking; machinery operation; offgassing from batteries and other construction materials; houskeeping activities; and the men themselves. Each of these sources gives rise to a complex array of contaminants. Many of these contaminants are harmless, but some are not; therefore, sampling and laboratory analysis methods must be available. This article describes these methods and also gives a brief description of the Navy's current program to develop a suitable in situ analyzer for submarines.

II. SAMPLING METHODS

There are two principal methods of sampling encapsulated environments: (1) air samples can be collected by various techniques; or (2) the atmosphere can be passed over some adsorbent material, e.g., activated charcoal, and the organic contaminants removed. The preferred method depends on the type of information desired. Air samples permit quantitative information to be obtained; but these data can be obtained only for a few contaminants whose concentrations dominate the mixture. In the second method, the contaminants are first concentrated on an adsorbent and then desorbed. This procedure permits large numbers of species to be identified. However, quantitative interpretation of these data depends on knowledge of the adsorption and desorption efficiency of each contaminant in the mixture. Moreover, many contaminants in an encapsulated environment either cannot be recovered from an adsorbent or may be altered by the adsorption-desorption techniques. Despite these shortcomings both sampling approaches of encapsulated air are used and will be described.

A. Air Sampling Methods

The simplest method for obtaining air samples of encapsulated environments is to open a previously evacuated flask in the environment at the appropriate time. We have used this method to obtain samples of submarine air for almost 20 years. We use 1.6-liter stainless-steel flasks with a bellows-type valve. The flasks are pressurized to approximately 60 psig with a carbon-vane or bellows pump, bled to atmospheric pressure, and repressurized several times to reduce contamination of the flask from previous samples. The flasks should always be pressurized so the analyst will know that the valve on the flask has not opened.

Although this method has the advantage of simplicity, it is not completely satisfactory. Frequently, contaminants from previous samples remain adsorbed on the walls of the flask. The valves sometimes leak between the time the sample is taken and when it is analyzed. The flasks are bulky and expensive.

A method which overcomes some of these objections has recently been developed by NRL. Clean, stainless-steel cans, similar in size to the hair spray containers, are commercially available [18] preevacuated. These inexpensive containers can be pressurized with a small hand operated, contaminant-free pump. After the contents of the can are analyzed, the container is discarded. Thus, contamination of the sample from the container is eliminated.

B. Adsorption Methods

Most of our analytical experience with encapsulated environments has been with samples desorbed from activated charcoal. The advantages and disadvantages of adsorption methods of sampling have been mentioned briefly above. These techniques now will be described in more detail.

Correct interpretation of data from charcoal sampling depends on knowledge of the adsorption and desorption efficiency of each individual component on charcoal; hence, while inferences can be drawn, no quantitative values can be assigned without specific adsorption-desorption data for the compounds identified in the contaminant mixture. Moreover, many contaminants in a submarine atmosphere either cannot be recovered from charcoal or may be chemically altered by the adsorption-desorption process. In addition to the charcoal sampling technique's being nonquantitative, the charcoal, itself, may contain a background of contaminants picked up elsewhere in its handling and storage. It is thus always desirable, but not always possible, to have an unexposed sample for a blank analysis. In spite of these shortcomings, the charcoal sampling method has the advantage that it is easily handled by a submarine crew without the need for auxiliary equipment.

It is convenient to take exposed charcoal samples from the ship's main charcoal filter, although such samples do not reflect an accurate picture of the contaminant profile in the ship at any particular time. The main charcoal filter may have been exposed up to 20 days; thus, at best, the charcoal sample is a time-integrated one. The sample gives no information as to where and when a particular contaminant appeared in the boat. It is also quite possible that some contaminants having short retention time on the adsorbent have been eluted from the filter. Thus, with increasing exposure time, the main filter charcoal is depleted of these volatile contaminants and enriched with contaminants having longer retention times.

Three methods of recovering adsorbed contaminants from charcoal have been employed at the Naval Research Laboratory: vacuum desorption [10], solvent extraction [19], and steam desorption [20,21]. We prefer the vacuum desorption method because no extraneous substances are added. The vacuum desorption apparatus used at NRL is all-glass and grease-free. The charcoal is desorbed under vacuum for 1-2 hr at room temperature, then gradually heated to about 300°C. The desorbate is retained in liquid-nitrogen-cooled traps and temperature fractioned after the initial collection.

Often it is important to know the contaminant makeup of the atmosphere at some particular time--for example, during an "atmosphere incident"--or at some particular location within the boat; and for this purpose, both whole air and special charcoal samples are taken. A sample on charcoal can be conveniently taken with a small analytical sampling device described elsewhere [22]. This device exposes a few grams of adsorbent to a known volume of air. The preconditioned charcoal is contained in metal canisters having a 2-in.-diameter and a 3/4 in. bed depth. These canisters are hermetically sealed both before and after exposure. A permanent record of the contaminant profile at any given time or location can thus be preserved for subsequent analysis.

Another method of sampling has been developed by the Anatole J. Sipin Co. (New York, N.Y.) under the sponsorship of the National Institute for Occupational Safety and Health. This method passes a known volume of air through a tube of solid adsorbent (we have used either charcoal or Tenax--a porous polymer material available from Applied Science Corp.). The pump is the size of a package of cigarettes (1 1/4 in. x 2 1/2 in. x 5 1/8 in.--weighing 11 oz) and can be easily carried in a shirt pocket. It is powered by a rechargeable nickel-cadmium battery which operates the pump for 8 hr at constant speed without recharging. The flow rate is adjustable and calibrated. Since the total air flow through the adsorbent is known, quantitative data on trace contaminants can be obtained by this sampling method if the adsorption-desorption properties of the contaminants on the solid sorbent are known.

III. ANALYTICAL METHODS

A variety of techniques are used to analyze the samples taken by the methods described in Section II: gas chromatography (GC); GC with microcoulometric detector; mass spectrometry (MS); infrared spectroscopy (IR); and the combinations of GC-IR and GC-MS. We have found GC-MS and GC with microcoulometric detector the more useful techniques. The Chemistry Division at NRL uses three GC-MS instruments for this purpose: a Hewlett-Packard 5930A GC-MS; a Perkin Elmer 900 GC coupled with a Varian dual ion source CH4 MS; and a Hewlett-Packard 810 GC coupled with a Finnigan Model 3000 MS. We have used many devices as the GC-MS interface. Some of the interfaces (frequently called molecular separators) which we have used are effusion tubes (made of porous glass, Teflon, and metal), membranes, and jet orifices. The purpose of the interface is to remove the carrier gas (usually helium or hydrogen) from the chromatographic effluent without removing any of the eluting components and to reduce the pressure of gas entering the mass spectrometer to a level which is not deleterious to the instrument. None of these interface devices has been completely satisfactory for every analysis of complex mixtures. Of the various types of molecular separators, the membrane type has been the most satisfactory for our routine analyses. When capillary columns, either the open-tubular type or the support-coated open-tubular type, are used, as much of the column effluent as possible is passed into the mass spectrometer. Therefore, a direct connection between the GC and MS is employed.

Our Hewlett-Packard 810 GC-Finnigan-3000 MS system has been modified to permit stop-flow techniques [22,24]. The flow from a first "survey" column can be interrupted at any time. This permits the material eluted under any peak, thought to represent a mixture, to be further resolved on a second column. Stop-flow techniques are used only for special analyses. Normally, only one column is used, and its output is passed either through a molecular separator or directly into the MS.

Halogenated hydrocarbons in submarine atmospheres are of particular concern to the Navy for three reasons: they have a generally undesirable physiological effect; they decompose in the catalytic burner to yield corrosive vapors; and certain decomposition reactions can yield extremely toxic products. NRL developed a special analytical technique using a GC with a microcoulometric detector to qualitatively and quantitatively determine halogenated hydrocarbons [25,26]. Since these procedures have been fully described elsewhere [22,23,25,26], only a brief résumé of the techniques will be given here.

The components of the NRL system include a MicroTek MT 220 gas chromatograph and a Dohrmann microcoulometer, the latter serving as a detector. An air sample is passed through a Porapak column maintained at 50°C in the GC. This concentrates the contaminants by separating them from the large volume of air with which they are diluted. The column is then temperature programmed with the multifunctional programmer of the MicroTek chromatograph. The effluent emerging from the chromatograph is next pyrolyzed at 1100°C, and the products of this decomposition are titrated by silver ions in the coulombic cell. The sensitivity of this technique ranges from 10 ppb to 10 ppm [25,

26]. The method is quantitative on an absolute basis, and hence requires no calibration standards. It is sensitive to all halogen-containing organic compounds except fluorides. Although most of NRL's experience has been with chlorides, this method has also been used to determine both bromine- and iodine-containing compounds [27].

IV. RESULTS

The sampling and analysis methods described in the previous section have been applied by NRL scientists to a wide variety of encapsulated environments. Samples of air from submarines [11-16], deepsea living quarters (*Sea Labs I* and *II*) [3-6], spacecrafts [10], and aviator's and diver's breathing gases have been analyzed. Not all of these analyses have been carried out in the same detail. It is impossible to identify all of the compounds in a submarine atmosphere. Indeed, this quantity of information is seldom needed. Occasionally in an emergency, it is necessary to identify as many of the compounds as possible, but this is an immense undertaking. This section will first describe the analytical results from various encapsulated environment studies and then the Navy's in situ analyzer program will be discussed.

A. Analyses of Submarine Atmospheres

Analyses of submarine atmospheres are both the easiest and hardest analytical problems we have undertaken. On the one hand, there are essentially no power or space restrictions on the analytical instruments used, and the crews are dedicated, intelligent, and well-trained. On the other hand, the mixture to be analyzed is probably the most complicated faced by an analyst. Saalfeld et al. [22,23] have presented data on the contaminants found in the *George Washington* and *Hammerhead*

TABLE 1

Contaminants in the U. S. S. *Hammerhead* Atmosphere

Fluorodichloromethane	Methyl acetate
Difluorodichloroethene	Ethyl acetate
Fluorotrichloromethane	Butyl acetate
Difluorochloromethane	Methyl alcohol
Difluorodichloromethane	Ethyl alcohol
Trifluorotrichloroethane	iso-Propyl alcohol
Ethylene	Butyl alcohol
Propylene	iso-Butyl alcohol
2,2,3-Trimethylbutane	sec-Butyl alcohol
Isoprene	Methyl chloroform
n-Pentane	Vinylidene chloride
iso-Pentane	Trichloroethylene
Pentene-1	Tetrachloroethylene
Cyclopentene	Chlorobenzene
Methylcyclopentane	Dichlorobenzene
n-Hexane	Acetaldehyde
Cyclohexane	Acetonitrile
Benzene	Ethyl nitrile
Toluene	Propyl nitrile
o-Xylene	Furan
Ethyl benzene	Dioxane
Indene	Ethylene glycol
C_9 aromatics	Methoxy acetic acid
C_{10} aromatics	Carbon disulfide
C_{14} aromatic (tertiary group)	Dimethyl sulfide
Acetone	Sulfur dioxide
Methyl ethyl ketone	Trimethylfluorosilane
Methyl isobutyl ketone	Silicones (methyl)

submarine atmospheres. These data were taken from a single temperature cut of a carbon desorbate and thus only begin to describe the complexity of the mixture present. Most of the contaminants in submarine atmospheres are hydrocarbons and require only characterization as to type rather than specific identification. A list of contaminants specifically identified in a typical submarine sample is given in Table 1. The compounds listed in this table represent two independent analyses and were detected in a volatile fraction of the desorbate. The total de-

TABLE 2

Summary of Total Hydrocarbon Concentrations Found by Carbon Sampling Method

Concentration (mg/m^3)	Ship
0-3	*George Washington*
2-10	*Abraham Lincoln*
0-22	*Theodore Roosevelt*
40-72	*Ethan Allen*
10-30	*Robert E. Lee*
0-13	*Robert E. Lee*
0-40	*Patrick Henry*
0-3	*Scamp*
0-2	*George Washington*
0-2	*Ethan Allen*
0-94	*Robert E. Lee*
0	*Theodore Roosevelt*
0	*Patrick Henry*
15-80	*Nautilus*
0	*Patrick Henry*
1-20	*Abraham Lincoln*
0-12	*Abraham Lincoln*
1-22	*Shark*
0-54	*George Washington*
0	*Ethan Allen*

sorbate, of course, comprises many additional compounds not listed in the table.

The NRL series of reports "The Present Status of Chemical Research in Atmosphere Purification and Control on Nuclear Power Submarines" has provided a wealth of background information on the contaminant profile of submarine atmospheres. Of particular interest is volume three [28] of this series which has several chapters dealing with the contaminant levels. Since it is difficult to obtain copies of these reports, some of the more important tables are presented here. Johnson [28], in Chapter 7, discusses the sampling and analysis of submarine organic contaminants. Table 2 shows data reported by Johnson for

TABLE 3

Analysis of Submarine Carbon Desorbate by Hydrocarbon Types[a]

	Content (Volume percent)		
	Aromatics	Middle fraction[b]	Saturates
Abraham Lincoln	29.6	7.3	63.1
George Washington	28.2	6.3	65.6
Snook	25.2	7.9	66.8
Robert E. Lee	23.1	3.4	73.6
Nautilus	23.7	5.9	70.4
Patrick Henry	37.0	6.5	56.5
Average	27.7	6.2	66.0

[a]Determined by the FiA Method (ASTM 1319-58).
[b]Usually referred to by petroleum chemists as olefins

TABLE 4

Compounds, Other Than Hydrocarbons, Quantitatively Identified in Nuclear Submarine Atmospheres

Material	Highest concentration found
Ammonia	2 ppm
Arsine	0.015 ppm
Carbon dioxide	0.8-1.1%
Carbon monoxide	20 ppm
Chlorine	1 ppm
Freon-12	50
Hydrogen fluoride	0.3 ppm
Hydrogen	0.35%
Methyl alcohol	6 ppm
Methyl chloroform	6 ppm
Monoethanolamine	1 ppm
Nitrogen	80%
Nitrogen dioxide	0.1 ppm
Nitrous oxide	27 ppm
Oxygen	20%
Ozone	0.5 ppm
Stibine	0.01 ppm
Vinylidene chloride	2 ppm
Water vapor	60 RH

TABLE 5

Hydrocarbons Quantitatively Identified in Trace Amounts in Nuclear Submarine Atmospheres

Material	Highest concentration normally found (ppm)
Acetylene	0.5
Benzene	0.01
Ethylbenzene	0.21
m,p-Ethyltoluene	0.34
Hydrocarbons, total[a] (excluding methane)	25[b]
Mestylene	0.29
Methane	118
iso-Propylbenzene	0.04
Pseudocumene	0.57
Toluene	0.25
m,p-Xylene	0.95
o-Xylene	0.32

[a]Total of all hydrocarbons, excluding methane but including aromatic and aliphatic and irrespective identification or measurement.
[b]Most values reported are less than 5 ppm.

samples obtained from nine different submarines, representing 20 different cruises. These data show submarines achieve reasonably good control of hydrocarbon contamination. The data in Table 2 also indicate that the hydrocarbon level in a given submarine may vary on different cruises. Note, for example, that the *Ethan Allen* cruises had hydrocarbon levels ranging from 40 to 72 mg/m^3 on one cruise to no measurable hydrocarbon level on another. Table 3 shows Johnson's data [28] for the analysis of hydrocarbon types from some submarine carbon desorbates. These data show that about 25-30% of the hydrocarbons found are the aromatic type. The consistency of the data from submarine to submarine is interesting and suggests a common hydrocarbon source.

TABLE 6

Compounds, Other Than Hydrocarbons, Qualitatively Identified in Trace Amounts in Nuclear Submarine Atmospheres

Material	
Acetaldehyde	Freon-114
Acetic Acid	Isopropyl alcohol
Acetone	Methyl ethyl ketone
Ethyl acetate	Methyl isobutyl ketone
Ethyl alcohol	Sulfur dioxide
Formaldehyde	

TABLE 7

Hydrocarbons Qualitatively Identified in Nuclear Submarine Atmospheres

Material	
iso-Butane	Ethylcyclohexane
n-Butane	Ethylcyclene
iso-Butane	o-Ethyltoluene
1-Butene	n-Heptane
2-Butene (trans)	n-Hexane
2-Butene (cis)	Isoprene
n-Butylbenzene	Methylcyclohexane
sec-Butylbenzene	n-Nonane
t-Butylbenzene	n-Octane
n-Decane	iso-Pentane
1-3-Dimethyl-5-ethylbenzene	n-Pentane
	Propylene
n-Dodecane	n-Undecane
Ethane	

In Chapter 8 of the same report [28], Carhart and Piatt have summarized the chemical constituents found in nuclear submarine atmospheres. These data are presented in Tables 4-7 and give some indication of the complex mixture of contaminants found in submarine atmospheres. Carhart and Piatt have related many of the identified constituents shown in these tables with those in cigarette smoke.

The value of the GC-microcoulometer technique for identifying the important halogen-containing organic materials is illustrated by comparison of the gas chromatogram measured with a flame ionization detector to one measured with the microcoulometer (Fig. 1). The output of the gas chromatograph was split, 10% to a flame ionization detector and 90% to a coulometric detector. The top trace depicts the flame ionization response and the bottom trace the coulometric response. The ease of observing trace amounts of chlorinated hydrocarbons in the presence of large concentrations of other hydrocarbons is readily apparent.

Saunders [29] has shown that a source of the chlorocarbons found in nuclear submarine atmospheres was commercial grade of methyl chloroform (1,1,1-trichloroethane) used as an adhesive cement solvent. He found some of these solvents contained as much as 30% by volume impurities.

B. Hyperbaric Environments

There have been numerous attempts to understand how man functions in high-pressure environments. These have included the *Sea Lab* and various other hyperbaric diving experiments. The high pres-

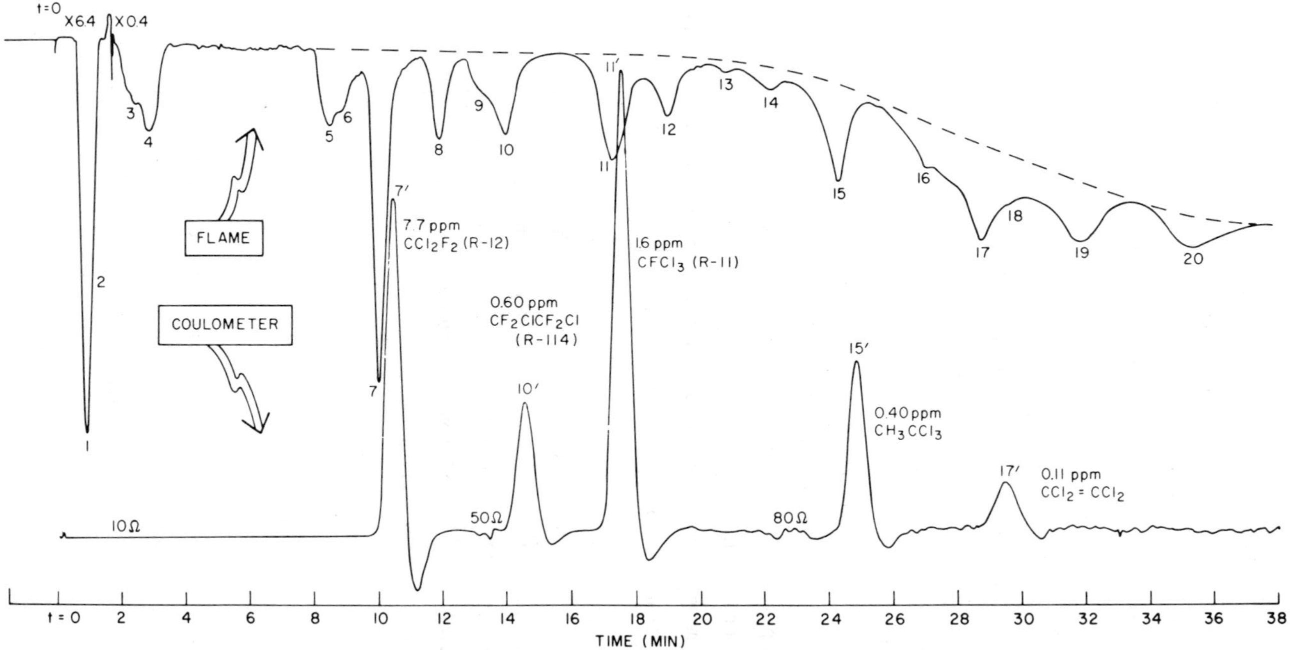

FIG. 1. Comparison of the microcoulometer and flame ionization detectors responses to a submarine carbon desorbate.

sures of these environments present benefits and drawbacks. For example, in most hyperbaric chambers the partial pressure of oxygen is too low to sustain combustion. Since smoking is impossible, the life support requirements are reduced. However, much smaller contaminant concentrations must be monitored. Saunders et al. [3-5] have reported the analyses of *Sea Labs I* and *II* atmospheres. In these experiments Navy divers lived in pressurized habitats on the ocean bottom. The atmosphere in *Sea Lab I* consisted of 4% oxygen, 16% nitrogen, and 80% helium at a total pressure of 110 psi. Four divers remained at a depth of 192 ft for 11 days. The contaminants of the *Sea Lab I* atmospheres were sampled with activated charcoal, then desorbed and analyzed at NRL. GC, IR and MS techniques were used for the analyses. A chromatogram of the vacuum desorbate mixture of the *Sea Lab I* charcoal showed 34 peaks. Mass spectral analysis of these peaks indicated most contained at least two constituents which were characterized as hydrocarbons. The major portion of these hydrocarbons had molecular weights greater than C_7. The overall composition of the mixture was similar to kerosene. The charcoal was saturated with the oil. The major source of contamination was traced to oil-treated dust filters [3].

The *Sea Lab II* experiments were more complex and consequently more elaborate analyses were carried out. The analytical results, shown in Table 8, provide an excellent opportunity to compare analyses from air samples with analyses of desorbates from activated charcoal.

Table 8 demonstrates that the two sampling methods complement each other well. Air samples provide quantitative data but the charcoal samples yield qualitative results, but in greater detail.

TABLE 8

Contaminants Detected in the Atmosphere of *Sea Lab II*

Contaminant	Detected in	
	Charcoal desorbate	Air Sample
Carbon monoxide	a	x
Carbon dioxide	x	x
Ethyl alcohol		x
Acetone	x	
Trichlorofluoromethane (refrigerant 11)	x	
Dichlorodifluoromethane (refrigerant 12)	x	
Difluorochloromethane (refrigerant 22)	x	
Methylene chloride	x	
Vinyl chloride	x	
Vinylidene chloride	x	
Methane	a	x
n-Butane	x	
n-Pentane	x	
n-Hexane	x	x
n-Heptane	x	x
n-Octane	b	x
n-Nonane	b	x
iso-Butane	x	
iso-Pentane	x	
iso-Hexane	x	
iso-Heptane	x	
3-Methylpentane	x	
2,2-Dimethylpentane	x	
2,3-Dimethylpentane	x	
2,4-Dimethylpentane	x	
Isoprene	x	
3,3-Dimethylpentene-1	x	
Cyclopentane	x	x
Cyclohexane	x	x
Methylcyclopentane	x	x
Methylcyclohexane	x	x
cis-1,2-Dimethylcyclopentane	x	
trans-1,2-Dimethylcyclopentane	x	
1,1,3-Trimethylcyclopentane	x	
Benzene	x	
Toluene	x	x
Ethyl benzene	b	x
o-Xylene	b	x
m-Xylene	b	x
p-Xylene	b	x

[a]Activated charcoal has a low affinity for these compounds.
[b]These compounds had an unreasonably long retention time for the column and conditions used.

TABLE 9

Analytical Results on Aviators' Compressed Breathing Oxygen

		Concentration (ppm)	
Contaminant	Formula	Cylinder no. 16	Cylinder no. 17
Carbon dioxide	CO_2	3.1	3.4
Nitrous oxide	N_2O	0.30	0.31
Acetylene	C_2H_2	0.20	0.15
Ethylene	C_2H_4	0.05	0.033
Propane	C_3H_8	Trace	0.056
iso-Butane	C_4H_{10}	0.012	0.012
n-Butane	C_4H_{10}	0.038	0.044
iso-Pentane	C_5H_{12}	0.025	0.012
n-Pentane	C_5H_{12}	0.015	0.036
C_6 hydrocarbons	C_6H_{14}	Trace of three compounds	Trace of six compounds

The most interesting feature of the data shown in Table 8 is the presence of seven cycloalkanes. Three of them--methylcyclopentane, cyclohexane, and methylcyclohexane--were found in significant concentrations. Considerable detective work was required to demonstrate that a cork adhesive was the source of these hydrocarbons [5].

One of the earliest studies at NRL on life support systems involved the analysis of contaminants in aviators' breathing oxygen. In this study, a combination of IR and GC techniques were used for the analyses of the contaminants collected and concentrated by an adsorption-desorption cycle. Activated charcoal was again used as the adsorbing material. Using this approach, Kagarise and Saunders [1] were able to detect and identify contaminants with concentrations as low as 0.5 ppb. The results of their analyses of two odorous oxygen samples are shown in Table 9.

Finally, Williams et al. [8] have developed guidelines for the field evaluation of small manned submersibles. Their chief analytical method is gas chromatography used in both conventional

TABLE 10

Analytical Results of the Contaminants Desorbed from the *Friendship VII* Carbon

Contaminant	Minimum concentration (ppm)
Vinylidene chloride	3
Benzene	3
Vinyl chloride	2
Methyl chloroform	0.5
Methylene chloride	0.4
p-Dioxane	0.3
Unidentified	0.3
Cyclohexane	0.2
Toluene	0.2
Methyl alcohol	0.2
Ethyl alcohol	0.05
Trichlorofluoromethane	0.05

and back flush modes. Their instrument, the total hydrocarbon analyzer, is now used as a backup analytical method on submarines.

C. Spacecraft Analysis

Saunders has used charcoal adsorbents to sample spacecraft atmospheres from the *Friendship*, *Mercury*, *Gemini*, and *Biosatellite* programs. In each of these studies he used GC, MS, IR, and GC-MS to identify the desorbed contaminants. Table 10 shows the results [30] obtained for the *Friendship VII* flight. Table 11 shows the data obtained for the *Mercury 10* capsule desorbate analysis. It is obvious that the spacecraft contaminant are not as complex as the mixtures found in submarine atmosphere. It is interesting to note the large number of chlorocarbons found in these spacecraft studies. Another very interesting finding was the detection of dichloroacetylene in one of the *Biosatellite* atmospheres by Saunders and Williams [31]. This study demonstrated the importance of removing chlorocarbons from an encapsulated environment and also presented an overview

TABLE 11

Analysis of *Mercury 10* Capsule Charcoal Desorbate

Compound (listed in order of decreasing concentration, except the last three)	Approximate minimum concentration in cabin atmosphere (ppm)
Freon-114	200
Ethylene dichloride	40
Toluene	20
n-Butyl alcohol	4
Trichlorofluoromethane	3
Vinyl chloride	3
Ethyl alcohol	3
m-Xylene	3
o-Xylene	1
Benzene	1
Methylchloroform	1
Methylene chloride	1
Trichloroethylene	1
Acetone	1
Methyl ethyl ketone	1
Methyl isopropyl ketone	1
Ethylene	1
n-Propyl alcohol	1
Acetaldehyde	1
Ethyl acetate	1
Freon-114, unsym.	1
Methyl alcohol	1
1,4-Dioxane	1
Carbon dioxide	
Water	
Oxygen	

of the problems the analyst faces in the detection of reactive constituents at concentration levels in the 0.1 ppm range.

D. In situ Analysis

As mentioned earlier, the present analytical tools available for in situ analysis of enclosed atmospheres are not very versatile. In particular, these instruments are unable to detect the presence of new and deleterious constituents as they appear in the atmosphere. In order to overcome these problems, the Navy has a program designed to develop new analytical instruments for use on board ships. The instruments being developed for submarine atmosphere analysis are known as the Central Atmosphere Monitor System or CAMS. The instrument resulting from the first phase of the CAMS program, the Mark I, is now in production. This unit consists of dual isotope fluorescence, nondispersive infrared spectrometer to monitor carbon monoxide and a fixed-collector mass spectrometer to monitor H_2 (m/e 2), H_2O (m/e 18), N_2 (m/e 28), O_2 (m/e 32), CO_2 (m/e 44), refrigerant 12 (m/e 85), refrigerant 11 (m/e 101), and refrigerant 114 (m/e 135). Obviously this system still suffers from a lack of versatility. There are presently two modern analytical instruments which have the needed versatility to solve this problem: the scanning infrared spectrometer and the scanning mass spectrometer. The former instrument, however, lacks the sensitivity to serve as the sole detector of atmosphere pollutants. The scanning mass spectrometer, on the other hand, has the sensitivity for trace analysis and, at least in theory, can be used to identify all the constituents in a submarine atmosphere. Scanning mass spectrometers have been considered for submarine atmosphere analyzers before, and have been rejected because of their fragile nature and the copious amount of data they generate. It was felt that these factors would make the instrument useless at sea. However, the emphasis of the space program accelerated mass spectrometry development and today the instruments are rugged, small, highly reliable, and the data they generate can be easily digested by a modern, medium-sized computer. Therefore, a mass spectrometer coupled to a computer is an ideal atmosphere analyzer for nuclear submarines. In addition, the computer can also be used to control other shipboard devices. The development of a computer-controlled analyzer is the second phase of the Navy's CAMS program, CAMS-II.

The heart of this instrument is a small, inexpensive mass spectrometer. Carbon monoxide is still monitored by an infrared spectrometer. The analytical scheme does not attempt to identify all the various hydrocarbons. Rather the hydrocarbons are characterized as aliphatic or aromatic by the following techniques. The ion current intensities at m/e 43, 57, 85, 99, and 113 are summed to give an indication of the aliphatic hydrocarbon concentration. The sum of the currents at m/e 91, 105, 119, 133, and 147 are an indication of aromatic hydrocarbon concentration in the submarine atmosphere and the m/e 78 ion current intensity is a measure of the benzene concentration in the atmosphere. Thus, important information, which is impossible to obtain with the present submarine analyzers, is readily available from the scanning mass spectrometer atmosphere analyzer. In CAMS-II, 12 preselected components will be continuously monitored and alarm systems installed. In addition, it will be possible to obtain a complete atmosphere analysis in less than 30 sec--a considerable improvement over present instrumentation. A complete mass spectrum will be recorded every hour on a magnetic disk (or some other device), and these records will provide the Navy with accurate information for future medical studies of the long-term effects of confined atmospheres on a sailor's health.

ACKNOWLEDGMENTS

We wish to thank our coworkers, particularly F. S. Thomas, F. W. Williams, R. S. Olfky, J. E. Johnson, and H. W. Carhart, at the Naval Research Laboratory who have contributed for many years to the analyses of encapsulated environments.

REFERENCES

1. R. E. Kagarise and R. A. Saunders, "The analysis of contaminants in aviators' breathing oxygen," *NRL Rept. 5554*, Oct. 1960.

2. F. W. Williams, "Source of contaminants in diving systems," *Proceedings: Purity Standards for Divers Breathing Gas Symposium*, July 8 and 9, 1970, Battelle Memorial Institute, Columbus, Ohio, Research Rept. 6-70, Chap. 6.

3. R. A. Saunders, "Atmospheric contamination in SEALAB I," *Proceedings of the Conference on Atmospheric Contamination in Confined Spaces*, AMRL-TR-65-230, Dec. 1965.

4. R. A. Saunders and R. H. Gammon, "The SEALAB II trace contaminant profile," NRL Rept. 6626, Dec. 1967.

5. R. A. Saunders, M. E. Umstead, W. D. Smith, and R. H. Gammon, "The atmospheric trace contaminant pattern of SEALAB II," *Proceedings of the 3rd Annual Conference on Atmospheric Contamination in Confined Spaces*, AMRL-TR-67-200, Dec. 1967.

6. M. E. Umstead, "Trace contaminants in the atmosphere of SEALAB II," *Sixth Annual Report, The Recent Status of Chemical Research in Atmospheric Purification and Control on Nuclear Power Submarines* (R. R. Miller and V. R. Piatt, Eds.), NRL Rept. 6722, June 1968, Chap. 9.

7. M. E. Umstead, "Trace contaminants in hyperbaric atmosphere," *Proceedings: Purity Standards for Divers Breathing Gas Symposium*, July 8 and 9, 1970, Battelle Memorial Institute, Columbus, Ohio. Research Rept. 6-70, Chap. 9.

8. F. W. Williams, M. E. Umstead, J. E. Johnson, and H. W. Carhart, "Guidelines for atmosphere evaluation of manned small submersibles and chambers," NRL Memo Rept. 2122, May 1970.

9. L. S. Young, "Summary of testing for trace contaminants in biosatellite III spacecraft atmosphere," NASA-TMS-62,004 (pending).

10. R. A. Sanders, "Analysis of spacecraft atmospheres," NRL Rept. 5316, Oct. 1962.

11. V. R. Piatt and E. A. Ramskill (Eds.), "Progress report: Chemical research in nuclear submarine atmosphere purification," NRL Rept. 7037, June 1970, and references contained therein.

12. J. C. White, "Atmospheric control in the true submarine," Report of NRL Progress, Dec. 1958, p. 1.

13. E. A. Ramskill, "Nuclear submarine habitability," SAE National Aeronautics Meeting, New York, 1961.

14. H. W. Carhart, "The maintenance of a habitable atmosphere," presented at the Underwater Technology Conference of the American Society of Mechanical Engineers, New London, Conn., May 5-7, 1965.

15. J. E. Johnson and J. K. Musick, "Studies of halogenated hydrocarbons. Part 1--Stability of fluorocarbons in air over hopcalite catalyst or stainless steel," NRL Rept. 6546, May 26, 1967.

16. J. E. Johnson and R. H. Gammon, "Halogenated hydrocarbons. Part 2--Decomposition of selected chlorinated hydrocarbons over hopcalite catalyst," NRL Rept. 6582, July 20, 1967.

17. H. F. Dymond and K. D. Kilburn, "The characterization of tobacco smoke by gas chromatography and a digital computer," *Gas Chromatography* (A. B. Littlewood, Ed.), Elsevier, New York, 1966. It should be noted that since cigarette smoking is permitted aboard submarines, many of the constituents found may be attributed to such smoking.

18. The sample cans used for the NRL study were No. 300709 aerosol cans with vacuum valves and were purchased from Barr Stalfolt, Baltimore, Maryland.

19. W. B. Moniz and C. F. Poranski, Jr., "NMR method for determining organic loading of charcoal," NRL Rept. 6566, June, 1967.

20. A. J. Chiantella, W. D. Smith, M. E. Umstead, and J. E. Johnson, *Amer. Ind. Hyg. Assoc. J., 27*, 186 (1966).

21. F. H. Nestler and W. D. Smith, "Submarine habitability--Atmosphere sampling and analysis," NRL Memo Rept. 866, Oct. 1958.

22. F. E. Saalfeld, F. W. Williams, and R. A. Saunders, *Amer. Lab.*, 8 (July 1971).

23. F. E. Saalfeld, R. A. Saunders, and F. W. Williams, "Analyses of Submarine Atmospheres," Report of NRL Progress, May 1971, p. 1.

24. R. D. Gribsly, E. J. Eisenbraum, D. V. Hertzler, and K. M. Piel, "Use of a stopped-flow MS-GLPC in the identification of reaction products from a Berch type reduction of napthalene," presented at the 15th Annual Conference on Mass Spectrometry, May 14-19, 1967, Denver, Colo., paper 123.

25. F. W. Williams and M. E. Umstead, *Anal. Chem., 40*, 2232 (1968).

26. F. W. Williams, M. E. Umstead, and J. E. Johnson, "Determination of trace quantities of halogenated hydrocarbons in gas samples, NRL Rept. 6964, Nov. 6, 1969.

27. D. M. Coulson, *Amer. Lab.*, 22 (May 1969).

28. H. W. Carhart and V. R. Piatt (Eds.), "Third Annual Progress Report--The Present Status of Chemical Research in Atmosphere Purification and Control on Nuclear-Powered Submarines," NRL Rept. 6053, Dec. 31, 1963.

29. R. A. Saunders, "Chlorinated hydrocarbons in closed environmental atmospheres," Chap. 2 in "The Present Status of Chemical Research in Atmospheric Purification on Control on Nuclear-Powered Submarines," A. L. Alexander and V. R. Piatt (Eds.), NRL Rept. 6491, Jan. 1967.

30. R. A. Saunders, unpublished data reported to NASA.

31. R. A. Saunders and F. W. Williams, "Another incident of dichloroacetylene contamination," Proceedings of the 5th Annual Conference on Atmospheric Contamination in Confined Spaces, AMRL-TR-69-130, Dec. 1969, p. 289.

Chapter 5

FIELD IONIZATION AND FIELD DESORPTION MASS SPECTROMETRY IN ANALYTICAL CHEMISTRY†

HANS D. BECKEY and HANS-ROLF SCHULTEN

Institut für Physikalische Chemie
der Universität Bonn
Bonn, Federal Republic of Germany

---

†This is an extensively modified and expanded version of an article that appeared in 1975 (see reference 32).

I. INTRODUCTION

Field ionization (FI) and field desorption (FD) mass spectrometry (MS) of organic substances today are considered standard techniques in analytical chemistry together with electron impact ionization (EI)-MS and chemical ionization (CI)-MS. Relatively well-developed FI/FD ion sources are produced in several different research laboratories. In addition combined sources for EI/FI/FD are commercially available.

For the analytical chemist the main area of application of FI and FD mass spectrometry is in the analysis of such substances which are sensitive toward thermal and/or electronic and vibrational excitation and therefore do not display molecular ions in their EI or CI mass spectra. FI mass spectrometry is also useful for the analysis of complex multicomponent mixtures. One more recent application of the FD technique is the direct pyrolysis of polymers on the high-field anode. Here the analysis of biopolymers and technical polymers is especially important. In this review article, however, we will not discuss the numerous applications of FI mass spectrometry in the various areas of basic research such as high-field surface chemistry (heterogeneous catalysis, ion-molecule reactions in the condensed phase) and the kinetics and decomposition mechanisms of organic ions. For reviews on these subjects see Refs. 1-6 and for a comparison of the EI, CI, and FI/FD methods, see Refs. 7 and 55.

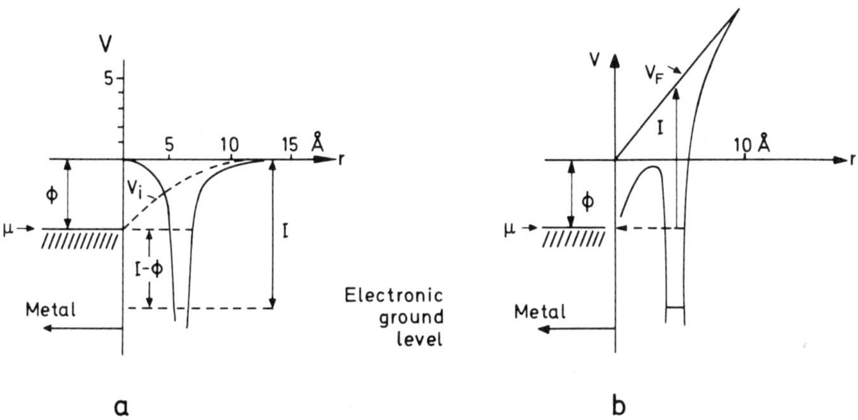

FIG. 1. Potential energy (V) of an electron of an atom near a metal surface: (a) Without external field, (b) with field. I= ionization energy; Φ = work function of the metal; V_i = image potential (dashed curve); V_F = external electric potential = e·F·r (F = field strength; μ = Fermi level).

II. PRINCIPLE OF FIELD IONIZATION

The phenomenon of FI of organic substances occurs at a field strength of several 10^7 V/cm. This is explained by means of the usual potential diagrams in Figures 1a and 1b.

Figure 1a shows the potential of a hydrogen atom in front of a clean and smooth *metal surface* without an external electric field. At $r < 0$ the electrons occupy the conductivity band of the metal up to the Fermi level μ; Φ represents the work function of the metal. At $r > 0$ the hydrogen atom is located a few angstroms in front of the metal surface in the vacuum. It is derived from Fig. 1b that ionization must take place at a minimal distance d_{min} so that the tunneling electron is raised to the Fermi level of the metal [8]:

$$d_{min} \simeq \frac{I - \Phi}{F} \tag{1}$$

For the organic analyst Eq. (1) is only of qualitative interest because the metal of the field anode is actually covered with a layer of an organic substance (e.g., needles of pyrocarbon). The dielectric constant of this layer is changed by the high electric field in a way which is theoretically difficult to understand. It is clear, however, that the levels of the electron acceptors of this layer play a significant role. The carbon needles are of practical importance because the onset of FI/FD occurs at lower field strengths as compared to metal needles. Equations for the qualitative calculation of the tunneling probability of electrons in simple atoms were given by Müller and Tsong [9] and by Gomer [10].

The total field ion current i^+ represents the product of the tunneling probability P and the particle supply S toward the emission centers and the electric charge e:

$$i^+ = P\ S\ e \tag{2}$$

The particle supply consists of two factors: (1) the particle supply from the gas phase $\dot{n}_g$; (2) the particle supply by surface diffusion along the shafts of the microneedles $\dot{n}_s$. Only the first factor can be calculated exactly for a single tip of defined geometry at zero field:

$$\dot{n}_o = \frac{{}^1n\bar{c}}{4}\ O \tag{3}$$

($\dot{n}$ is the impact rate, ${}^1n$ is the particle number/cm^3, $\bar{c}$ is the average molecular velocity, O is the effective tip surface).

Müller [12] was the first to point out that the particle supply calculated according to Eq. (3) can be enhanced considerably by an extremely high electric field. The molecules in the regions of high field strengths can be increased by polarization forces; therefore an "enhancement factor" σ is introduced so that $\dot{n}_g = \dot{n}_o\sigma$,

$$\sigma = 1 + \frac{2}{3}\,\frac{V_{(F)}}{kT} \tag{4}$$

where $V_{(F)}$ represents the polarization energy:

$$V_{(F)} = \frac{\alpha F^2}{2} + \bar{\mu}F \qquad (5)$$

where α is the polarizability and μ is the averaged dipole moment of the sample molecule in the direction of the field. The second factor of the particle supply derived from surface diffusion along the shafts of the microneedles is theoretically difficult to determine [8]. Under the usual conditions of FI of organic molecules this factor is considerably larger than that produced by direct impact of particles from the gas phase on the ionization centers [13].

Southon and Brandon [14] have measured the current-voltage curve for the field ionization of helium on tungsten tips (Fig. 2).

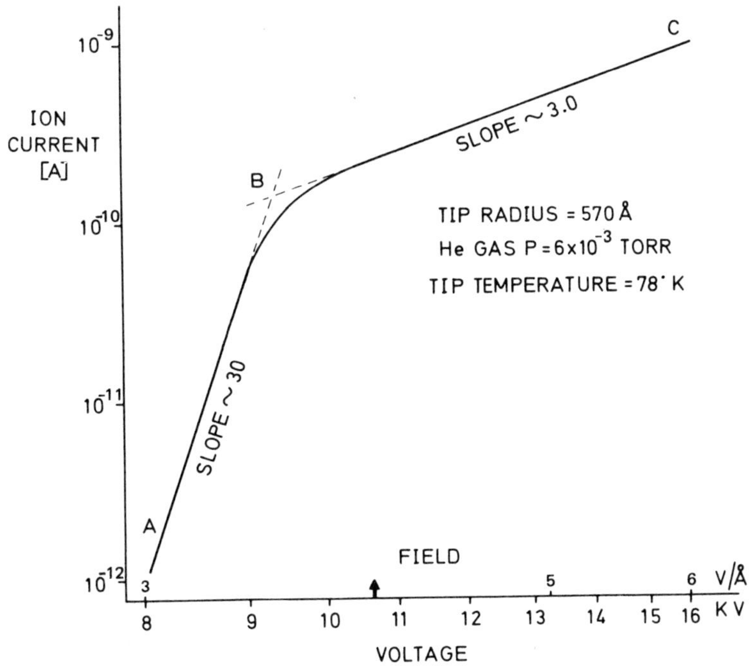

FIG. 2. FI current-voltage curve for helium at a tungsten tip [14].

Near the "threshold field strength" below which the current is unmeasurably small, the field ion current increases extremely rapidly:

$$i^+ \sim F^{30}$$

Above the point of inflection B, which is obtained by extrapolation, the ion current increases rather weakly with the field strength:

$$i^+ \sim F^2 \text{ to } F^3$$

From this (and from the theory of field ionization) it must be concluded that the field ionization probability at point B is equal to one. This means that a further increase of the ion current can only be effected by the enhanced supply rate of neutral molecules due to increased polarization.

In practice the analytical chemist does not use single tips but activated emitters with an extremely large number of ionization centers. In this case the point of inflection B of the current-voltage curve cannot be formed by extrapolation of two linear ranges of curves. Metzinger and Beckey [13] have attempted to calculate theoretically the current-voltage curve of activated emitters. Although these calculations are doubtlessly insufficient, it is immediately evident that due to the numerous needles with different radii of curvature and due to the distribution of field strengths the resulting current-voltage curve for an activated emitter is much smoother than that for a single tip (Fig. 3).

Another important observation when using activated emitters is the fact that the field dissociation of organic molecules is much less dependent on the applied voltage than when a single tip is used. Here, too, measurements have been made, especially on n-heptane [15] and water [16]. Based on these measurements, it is possible to characterize qualitatively the average field strength on activated emitters; for this purpose ratios

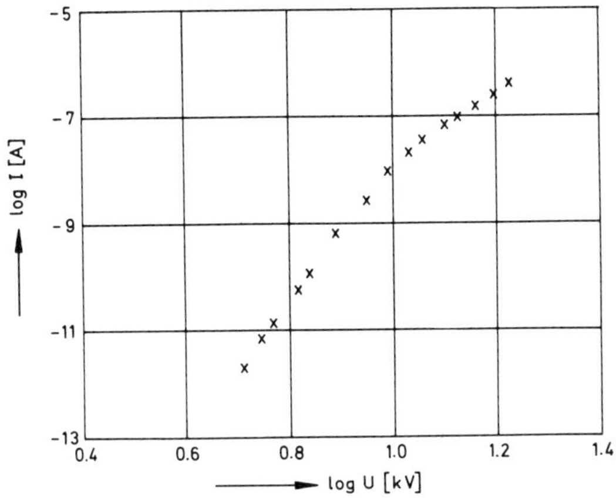

FIG. 3. FI current-voltage curve for n-heptane on a 2.5 μm platinum Wollaston wire covered with microneedles of organic material [13].

of m/e = 29 (for $[C_2H_5]^+$) to m/e = 100 (for the n-heptane molecular ion) are measured, or better still the ratios of m/e = 19 (for $[H_3O]^+$) and m/e = 18 (for $[H_2O]^{+\cdot}$).

III. PRINCIPLE OF FIELD DESORPTION

Only at very high field strengths can organic molecules be ionized in the gas phase several angstroms *in front* of the surface of the field anode. With the activated field anodes usually employed for analytical purposes and at field strengths of a few tenths of volts per angstrom, the large majority of the molecules are field ionized in the adsorbed state. If the ionized molecules are desorbed (which is not necessarily the case [17]), the term *field desorption* is used. Müller [18] and Inghram and Gomer [19] were the first to carry out investigations on the field desorption of metal atoms and of small molecules on metal tips. The field desorption of an atom with high ionization potential is shown in the potential curves of Fig. 4, according to Gomer [8]. This diagram is only valid for metallic field anodes. Figure 4a shows the potential curves

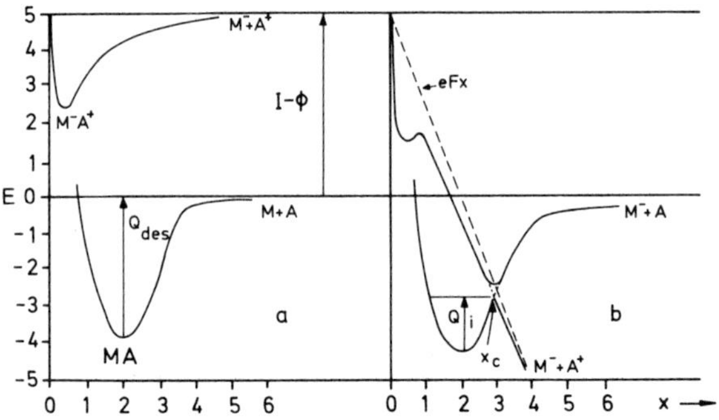

FIG. 4. (a) Potential curves for the interaction of an atom (A) and an atomic ion $[A]^+$, respectively, with a solid metal surface (M) in the field free space (Φ = work function of the metallic emitter). (b) Potential curves of the same system in the presence of a high electric field.

for the interaction of an adsorbed neutral or ionized atom with a *clean* metal surface M [8]. Here it is assumed that the ioniation energy I of the atom A is substantially greater than the work function Φ of the metal. For the transfer of the adsorbed neutral atom into the ionized state of the free atom $[A]^+$, the energy, $E_{des} = Q_{des} + I - \Phi$, is required (several eV). By superimposing a high electric field (~1 V/Å) the potential curves represented in Fig. 4b are produced. For the formation of the $[A]^+$ ion, only the considerably smaller energy Q_i is required. At extremely high field strengths the atom can be desorbed in the ionized state without any additional thermal energy.

The detailed theory of the field desorption of atoms on pure metal surfaces is very complicated if all polarization terms, appearance potentials, resonance energies, etc., are to be taken into consideration. Moreover, the opinions of different authors about the theory of field desorption and of field evaporation are controversial. A more thorough treatment of these problems can be found in the book by Müller and Tsong [9].

A quantitative theory of field desorption of large organic molecules on the microneedles of graphitized carbon does not

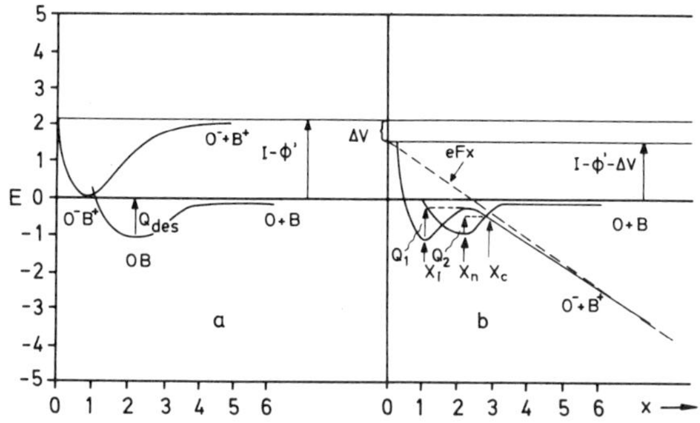

FIG. 5. (a) Potential curves for the interaction of a molecule (B) and a molecular ion $[B]^+$, with a surface (O) of solid organic material, in the field free space. (Φ' = work function of the organic emitter). (b) Potential curves for the same system in the presence of a high electric field. For desorption of an ion adsorbed at x_i the field desorption energy Q_1 and for desorption of a neutral molecule adsorbed at x_n the energy Q_2 must be supplied (x_c = intersection of the potential curves for the neutral and the ionic state of the molecule). A separation of the potential curves because of the "noncrossing rule" is not shown in Fig. 5b, in contrast to Fig. 4b.

exist at this time. The needles consist partly of organic polymers and partly of regular graphitized regions [20]. During the field desorption of organic molecules from this type of field anode, such complicated effects occur that only a qualitative explanation for the observed phenomena will be given in Figs. 5a and 5b†. The following effects, among others, must be considered:

Only a portion of the organic molecules is chemisorbed at the surface of the field anode. The molecules which are easily field desorbed are bound to the surface by van der Waals and dipole forces. The potential trough for neutral molecules is therefore considerably more flat than shown in Fig. 4a.

---

†The description of field desorption of organic molecules from carbon needles in terms of potential curves is applicable only to very low coverage of the carbon surface (i.e., maximum a monolayer adsorption).

The potential curve for the adsorbed ion in the high electric field is shifted to lower potentials by ΔV. ΔV is composed of several parts. The largest part can be attributed to the potential drop in the microneedles relative to the metallic carrier wire. Another part is due to the penetration of the electric field into the microneedles.

The electron work function can become greater by several eV with organic emitters because of their low-lying localized energy levels as compared to metallic emitters.

Owing to the preceding effects, the intersection of the potential curves may occur for the neutral and ionized state of the molecule at one or several distances from the surface of the microneedles (Fig. 5b). Fundamentally one can distinguish between the following cases according to the position of the points of intersection.

1. The organic molecule is present as an ion in the adsorbed state. It must then be thermally activated for desorption.
2. The molecule is adsorbed in the neutral state. It must then also be activated for transfer into the ionized state and subsequent desorption. The activation energy in both cases is small (of the order of 0.5 eV) so that only slight heating of the field anode leads to field desorption.
3. The molecules, ionized or neutral, are adsorbed relatively far from the ionization centers, i.e., in those regions of the field anode where the field strength is not sufficient for field desorption. In this case the molecules must migrate to the ionization centers by diffusion.
4. The possibility exists that molecules can be field desorbed in the ionization centers without thermal activation. In that case, the temperature-dependent transport processes alone are responsible for the desorption rate.

Whereas the description of FD in terms of potential curves is still valid for very small coverage of the surface, in all cases of analytical interest thick layers of the organic samples on the shafts of the carbon needles are realized.

Two different cases must then be distinguished.

Case (a). Nonelectrolytic samples. These are pure organic samples which are neither salts nor acids. Pure means here that within the different alkali ions it is the lithium ion which has the highest ion affinity. The solvated ions in the sample layers metal cations (mainly alkali ions) must be excluded even in spurious concentrations. Let us take a highly branched solid hydrocarbon as an example. The sample is deposited mainly on the shafts and in the gaps between the carbon dendrites of the activated emitter. The field strength is very low at these sites. The sample molecules can then be evaporated from this large reservoir of neutrals so that a fraction of them will be field ionized at the fine tips of the microdendrites. The sublimation pressure of such samples can be relatively high as compared with nonvolatile samples containing many polar groups. In the latter case the sublimation pressure is very low, and it can be assumed that the sample must diffuse on the shanks of the dendrites to the ionizing tips, or more probably, it must melt, and a flow of the liquid (or possibly a quasi-liquid [137]) sample to the tips occurs. The latter possibility is supported by observation of a "best anode temperature" BAT (see Sec. IV.D).

For such samples having a number of polar groups (but no acidic groups), i.e., in "nonelectrolytic" samples [201], field desorption can be regarded as field ionization in a thick layer on the fine dendrite tips. The layer thickness exceeds d_{min} as defined by Eq. (1). The electric field strength for FD of pure samples of case (a) must be in order of 10^7 V/cm.

Case (b). Electrolytic samples. (I.e., compounds having acidic groups, or salts.) Röllgen et al. [140] and Heinen et al. [141] have discussed the mechanism for FD of such samples. In addition to the process of ion formation within the sample layer the processes at the phase boundaries, i.e., the emitter/sample and the sample/gas interfaces must be taken into account. The ionization process within the layer is a "cationization" reaction as discussed by Röllgen and Schulten (see Ref. 39, for example). Cationization is most efficient with alkali ions, and must be desolvated for field desorption into the gas phase. An electric field of the order of 10^6 V/cm is necessary to overcome the desolvation energy. This is an order of magnitude less than in Case (a). Therefore it is not necessary to use activated emitters for FD of electrolytic samples but a 10 μm tungsten wire having a natural roughness from the wire drawing process is sufficient. On the other hand, 10^6 V/cm is still a very high field so that the designation of this process by "Field Desorption" is justified. A field-independent ion desorption process at a few hundred degrees Celsius as postulated by Holland et al. [137] does not exist (see Ref. 142).

IV. EXPERIMENTAL TECHNIQUES

A. Field Anodes for FI and FD Mass Spectrometry

For the production of extremely high field strengths very fine metal tips, metal blades, or thin wires are used which are covered with numerous organic or metallic microneedles.

1. Metal Tips

The first FI mass spectra were recorded by Inghram and Gomer [19] as well as by Müller and Bahadur [22] with tungsten tips of about 1000 Å radius of curvature. The advantage of the single metal tip is that the field strength can be raised very high and can be varied within great limits. A further advantage is that its geometry can be defined rather exactly. Thus, the use of single tips is well suited for the examination of the basic phenomena of field ionization. It is possible, for example, to describe catalytic processes by varying the point metal (or semiconductor) and to compare theoretical and experimental results in field ionization by calculating the field strength. In addition it is possible by the use of metal tips to routinely reproduce surface processes in the field ion microscope (FIM) and to correlate the results with those obtained in the mass spectrometer. In the "atom probe" developed by Müller [11], one can even determine mass spectrometrically the mass of a single atom previously made visible in the FIM, all in the same apparatus. However, it is not the aim of this review article, written primarily for the organic chemist, to discuss the interesting results obtained with this technique.

For analytical applications (apart from metallurgical investigations) the single point has the disadvantage that the emitting area is too small for reproducible measurement of the field ion current. With "activated" field anodes, however, there are an extremely large number of ionization centers so that the field ion current is sufficiently intense when averaged over the emission fluctuations of the individual centers. The "multipoint emitter" represents a compromise between a single point and activated emitters. It was developed by a research group at the Stanford Research Institute [23]. It is an array with a very regular arrangement of about 2500 single metal points which are produced by a special vaporization technique incorporating semiconductor technology. The dimensions of the points are: 40 μm in length, a distance of 25 μm between the needle points, and a radius of curvature of 1000 Å. The distance between the points grown in the array and the grid-shaped counterelectrode is only about 50 μm. One tip is located in the center of each mesh of the counterelectrode. A comparison of the sensitivities of a multipoint source operated in the FI mode and a typical activated wire operated in the FI mode yielded a sensitivity with cholesterol which was about 10-100 times greater for the activated wire (in C/μg of the mass resolved ion beam). It has been shown in pictures taken with a scanning electron microscope that fine needles grow on the multipoints after some time of operation in the presence of organic substances (i.e., an activation of the points has taken place). Under these circumstances it would be more efficient to use activated wire emitters from the start since the linear geometry of a wire is better suited to the slit geometry of magnetic mass spectrometers of conventional design than the array arrangement of the multipoint emitter. There might be a certain advantage in the use of the multipoint emitter when coupled with a quadrupole mass spectrometer, since the ion optics here show a spherical symmetry. Smaller retardation might also result with a certain enhancement of the transmission. Investigations of Gierlich et al. [123] showed that a reduction of the retardation results only in an unimportant enhancement of the transmission of a quadrupole mass spectrometer. This is presumably because of the extremely great divergence of angles of the emitted field ions. Systematic comparisons about these questions do not yet exist. It is important for the comparison of the different emitter types that the multipoint emitter is obviously not suited for field desorption. By the deposition of solid organic substances on the multipoints and due to the small distance to the counterelectrode, arcing can easily occur and destroy the emitter. Moreover, neither the emitter dipping nor the syringe techniques (see Sec. C.1) can be applied to this type of multipoint field ionizers.

2. Metal Blades

Sharp metal blades (e.g., razor blades or foils) have often been used as field anodes in FI mass spectrometry. If they are not activated they can be used--just like tips--in order to explain questions about basic phenomena. Compared to single tips they have the advantage of a greater emission area, but the disadvantage of a less well-defined geometry. With new blades questions about heterogeneous catalysis and about the kinetics of the fast decomposition of organic ions can be explored. From the kinetics it is often possible to draw conclusions about the mechanism of ion decomposition [6]. For field *desorption* mass spectrometry, however, activated thin wires are best. The geometry of thin wires is better defined and the distance to the counterelectrode may be greater by about one order of magnitude than that of blades, which helps to avoid arcing. The current necessary for heating wire anodes is several orders of magnitude smaller than that needed for heating metal blades. Thus the field desorption ion source in which an activated thin wire is used as field anode stays a "cold" ion source, which has advantages as we shall demonstrate later.

3. Thin Wires

Presently, 10 μm tungsten wires are used almost exclusively for analytical purposes in FD mass spectrometry†. Tungsten is preferred because of its very high resistance to mechanical stress and because very stable tungsten carbide nuclei are formed on the surface at the start of the activation procedure. The use of thicker wires is not advisable, because the so-called "basic field strength" (i.e., the field strength of the smooth, non-activated wire) would then be too small. It is impractical to use much thinner wires, since the probability of rupture decreases with $1/r^2$. Moreover, it is also not useful, since field desorption requires that the length of the needles grown during activation be about 20-50 μm for sufficient sample deposition; a reduction of the wire diameter below 10 μm would not contribute much to the enhancement of the field strength. For FI, 5 μm wires with relatively short microneedles are preferred (∿0.5-1 μm) [24].

B. Methods of Activation

Because of the preference for 10 μm wires for analytical purposes, we shall only discuss the activation of these wires. It must be pointed out, however, that single metal points and blades can also be activated. Activation is the formation of needles on a field anode by high-field polymerization or field pyrolysis of organic substances. After a systematic investigation of a great number of organic substances it was found that the activation is more efficient if the ionization potential is low and

†The 10 μm tungsten wires of the Lumalampan, Co., Stockholm 20, Sweden or Radium GmbH, Wipperfürth, BRD, are best.

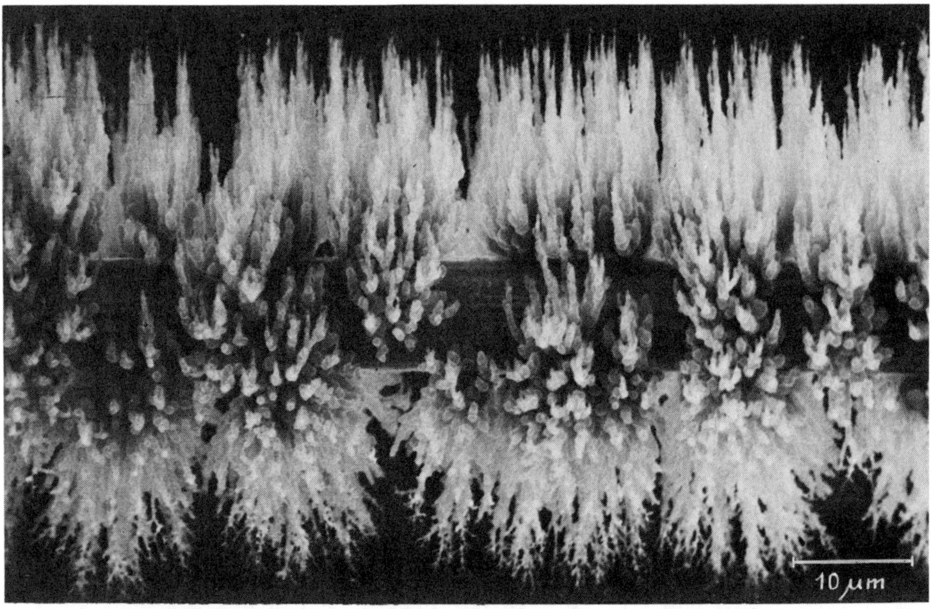

FIG. 6. Reproduction of a 10 μm tungsten wire activated with benzonitrile vapor at 1200°C., taken with a scanning electron microscope. The length of the carbon microneedles is about 20 μm [25].

the dipole moment of the substance is great. It should, moreover, contain a number of conjugated double bonds. Considering these aspects benzonitrile was found to be the most ideal activation substance. One differentiates between activation at room temperature (RT) and at high temperature (HT). The latter has the following advantages: the needles are of higher mechanical stability; they are more resistant to thermal stress (up to about 2000°C); and they display a greater chemical resistance toward corrosive substances. *The salient feature for the application of FD-MS is the optimal quality of the activated emitters.*

The mechanism of the formation and structure of the HT needles has been partially explained by experiments with the scanning or transmission microscope and by electron diffraction [20]. Figure 6 shows a 10 μm tungsten wire activated with benzonitrile at high temperature [25]. The following mechanism is proposed for needle formation: Without any applied high electric field but in the presence of benzonitrile nuclei of tungsten carbide are first formed at approximately 1200°C along the wire ridges which formed while drawing out the tungsten during the course of wire production. Subsequently, the needles are grown at a field strength of several tenths of volts per angstrom by deposition of pyrocarbon.

More than two thousand 10 μm tungsten wires have been activated with a success rate of more than 90% by a procedure developed in the laboratories of the authors [25,26]. This procedure in which 10 field anodes are activated simultaneously can be described in detail as follows: The activation cells are filled with benzonitrile vapor at about 10^{-2} torr (Fig. 7 shows a single cell). The vapor pressure is measured with a heat-conductivity manometer. A heating current of about 48-50 mA is passed through the wire without high voltage by variation of resistor R_2 which is in series with the resistance R_1 of the anode wire

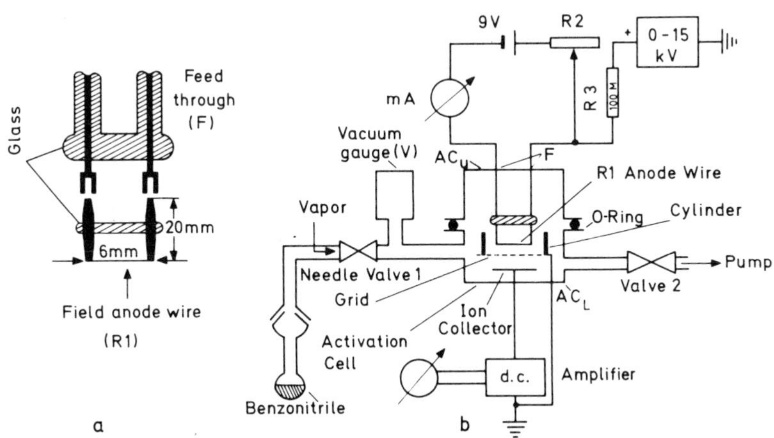

FIG. 7. (a) Carrier of a field anode wire. (b) Cell for activation of a single wire (schematically). AC_u = upper part, AC_L lower part of the activation cell. F = electrical feedthroughs. A glass window at the end of a side tube of the activation cell for pyrometric measurement of the anode wire temperature is not shown in the figure.

(or wires). The wire is adjusted to 1200°C with the help of an optical pyrometer. In the following minutes the heating current decreases owing to the reaction of the tungsten with the benzonitrile to form tungsten carbide nuclei. In order to avoid irregularities during the formation of nuclei, Barofsky and Barofsky [27] applied an oxidation-reduction procedure prior to activation in order to roughen the surface of the tungsten wire. This should facilitate the formation of nuclei for further activation. The present authors, however, found that this is unnecessary if a very clean tungsten wire is used for activation.

The temperature of the tungsten wire increases to about 1400°C during the course of the formation of carbide. This temperature is too high for uniform activation. In the middle of the wire, where the temperature is highest, only very short needles are formed. By reducing the heating current to about 40 mA the temperature is decreased again to about 1200°C.

After the establishment of a stationary temperature for a few minutes, the high voltage of about 10-12 kV is switched on. Growth of the needles now proceeds continuously. This can be monitored by the measurement of the emitted benzonitrile field ion current. The ion current which is measured on the counter-electrode increases to about a maximum of 3 μA. The pressure of the benzonitrile vapor should be regulated in such a way that it stays just below that value which produces an electrical discharge. After about 1 hr, the high voltage is decreased to about 5-6 kV. This is because the needles grow longer at this voltage relative to the higher voltage. Moreover, one can also increase the benzonitrile pressure by about a factor of 2 at the lower voltage. The rate of activation is thus increased. The total activation time for growth of needles of about 15 μm length is about 4 hr and for the needles of about 30 μm length is about 7 hr. After about 3 hr of activation the heating current of the wire must be increased by about 8 mA/hr because the temperature of the tips of the microneedles decreases below a value of 1200°C.

The distance between the field anode and the counterelectrode is kept constant at about 2 mm during the activation process. With smaller distances the needles grow preferentially in the direction of the counterelectrode. Activation can also be carried out automatically, as shown by Winkler et al. [28], by keeping the temperature constant with the use of a photocell.

The HT-activated carbon emitter can be used for the measurement of several samples, unless one sample is so nonvolatile that the required field desorption temperature is extremely high. In this case (e.g., with oligosaccharides) a nonvolatile, tarry deposit is formed between the microneedles which drastically reduces the emitter quality, and thus the emitter can be used only for the measurement of one sample. If the emitter quality is reduced only slightly by measurement of several samples, it can be improved again by repeating the activation procedure for short periods (∿5-10 min) in the activation device. This reactivation of FD emitters has been described explicitly in Ref. 90. Normally the carbon emitters afterward can be used up to 10 times. A review of other methods of FD emitter activation is given in Ref. 21.

C. Field Desorption Techniques

The basic phenomenon of field desorption was described in Sec. II. In the following discussion, the word "field desorption" will be used synonymously for a certain technique of sample supply to a field anode. (The designation FD-MS is used correspondingly.)

1. *Sample Supply Techniques*

a. Emitter dipping technique. A solid organic sample is dissolved in a suitable solvent (if such a solvent cannot be found one produces a suspension of the substance [32]), and then the activated wire is dipped into a small amount of this solution. One needs only a few microliters of solution, which is put into a suitable glass tube (Fig. 8a). After taking the emitter out of the dipping tube, part of the solution remains on the microneedles (the free volume between microneedles of about 30 μm length is about 10^{-2} μl). Part of the solvent will evaporate at atmospheric pressure and the rest after introducing the emitter through the vacuum lock into the FD source (Fig. 9). With this emitter dipping technique [29] one can measure solutions of concentrations down to about 1 μg/μl at a mass resolution of m/Δm approximately 1000 (corresponding to 10^{-2} - 10^{-6} molar solutions at MW = 100).

The emitter dipping technique is useful if a sufficient amount of solution (several microliters) is available and if the concentration is high enough. Maurer et al. [30] have shown that the ion currents of the antibiotics, cytarabine and tauromycine, are proportional to the concentrations in the quoted range within about ±20%. If the volume of the solution is only

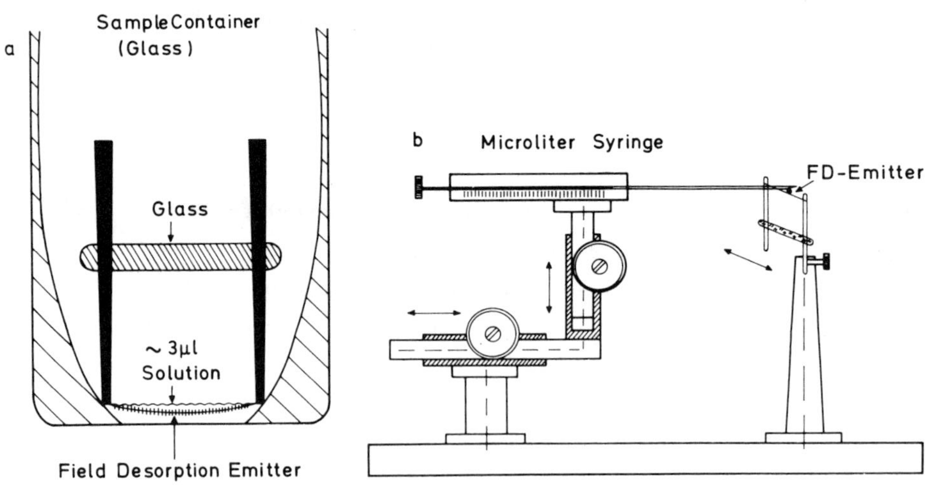

FIG. 8. (a) Emitter dipping technique. (b) Micromanipulator for the syringe technique.

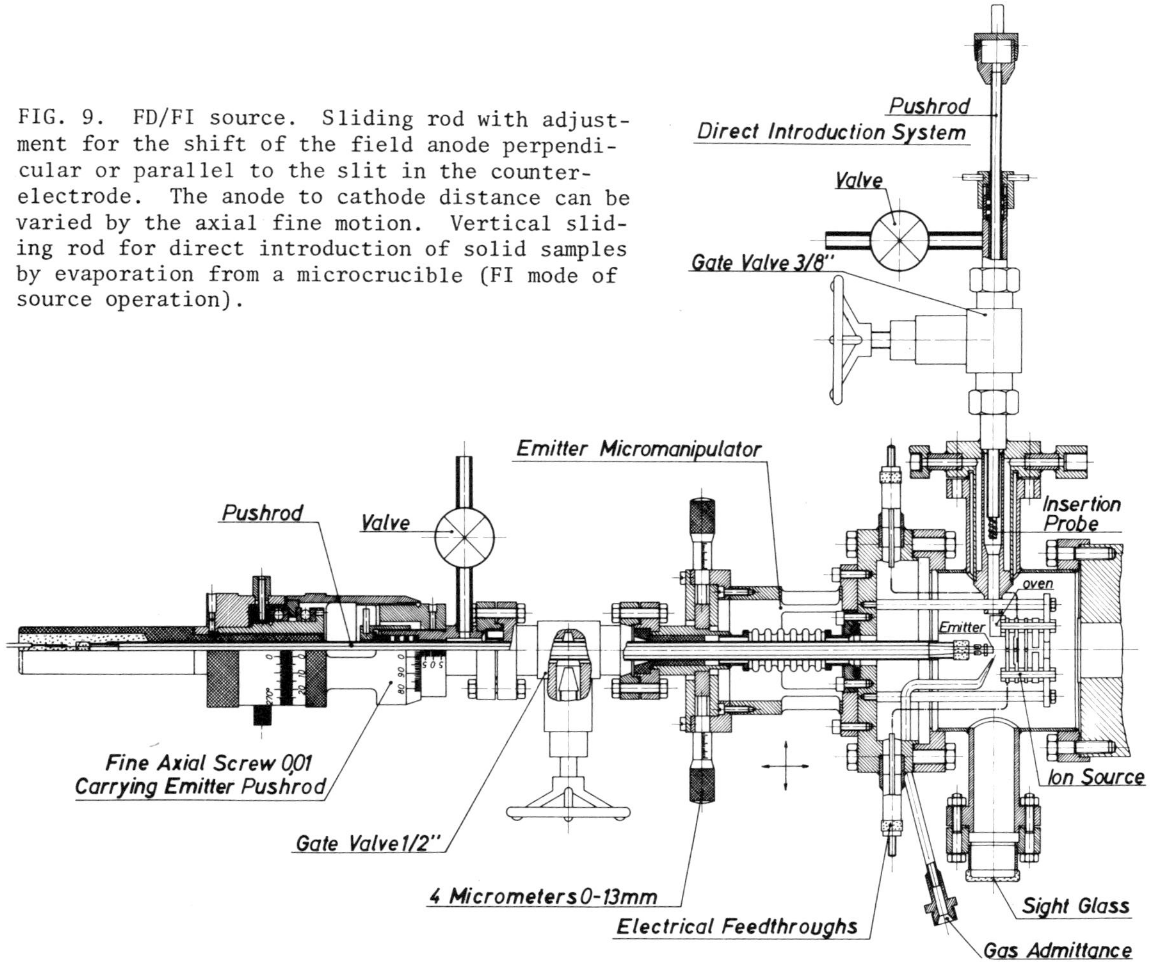

FIG. 9. FD/FI source. Sliding rod with adjustment for the shift of the field anode perpendicular or parallel to the slit in the counter-electrode. The anode to cathode distance can be varied by the axial fine motion. Vertical sliding rod for direct introduction of solid samples by evaporation from a microcrucible (FI mode of source operation).

of the order of 1 μl or less, or if the *absolute amount* of the substance deposited on the field anode has to be determined (e.g., for measurement of the sensitivity in coulombs per microgram of the adsorbed substance), the emitter dipping technique must be replaced by the syringe technique [31].

b. Syringe technique. A wetting agent such as dodecyclbenzylsulfonate is used if a substance to be analyzed is soluble only in a solvent that will not wet the carbon needles [32]. The emitter carrier is clamped rigidly, and a microliter syringe is fixed in a micromanipulator which can be moved in three orthogonal directions in space (Fig. 8b). The syringe must approach the emitter very carefully. A droplet of about 1 μl is pushed out of a 10 μl syringe, which is moved at an angle of about 45° with respect to the vertical direction, until a droplet of the solution containing the wetting agent distributes spontaneously over the emitter. This procedure is repeated with the solution of the sample. According to the method of Schulten [32], the emitter is heated with a current of about 40 mA so that the heat of evaporation of the solvent is just compensated. The emitter is continuously watched through a stereomicroscope. It is moved horizontally repeatedly until the droplet is evaporated. If larger volumes (up to 10 μl) are to be placed on the wire, one pushes a new droplet of 1 μl from the syringe. This can be repeated about 10 times during a few minutes. In this way standard solutions of about 10^{-8} M can be well measured with the FD technique; however, it is not recommended for the study of solutions with molarities much below 10^{-8} because then the concentrations of the impurities might be in the same order of magnitude as those of the sample. The heating current of the wire by a dc source must be adjusted by a variable resistance from 0 to 70 mA. In the microscope one can observe the evaporation of the solvent and one regulates the temperature so that it is only slightly above ambient temperature. This must be done with great care because after quantitative evaporation of the droplet, the wire temperature increases quickly, and this can lead to destruction of the sample or the emitter. When the droplet has evaporated one has to switch off the heating current immediately or a new droplet has to be put on the wire. If a solvent has to be used which evaporates very fast, the emitter can be cooled by solid CO_2 [33] so that the solution first freezes out and then the solvent is slowly evaporated.

2. Production of FD Mass Spectra

The emitter coated with the substance by either the emitter dipping or the syringe technique is introduced into the ion source of the mass spectrometer by means of a sliding rod through a vacuum lock. The rest of the solvent evaporates quickly in the high vacuum of the ion source. Before taking an FD mass spectrum the emitter must be adjusted mechanically and the ion beam must be adjusted electrically. This is usually done by introducing acetone via the gas-inlet system and producing an FI mass spectrum. The field anode is adjusted mechanically by a micromanipulator (with mass spectrometers modified by the authors of the types Varian MAT CH4, CH5, AEI MS9, and CEC 21-110B) or brought into a fixed, exactly defined geometrical position in the ion source (mass spectrometers of the types Varian MAT CH5, 711, 731). The quality of the emitter can be judged from the intensity of the acetone FI current. After optimization of the acetone ion current the supply of acetone is interrupted. The field anode is supplied with a positive potential usually of 3-10 kV with respect to earth potential, and the counterelectrode at a distance of about 2 mm is supplied with a negative potential of several kilovolts, so that the total voltage between the field anode and the counterelectrode is approximately 10-12 kV. Now a heating current of several milliamperes is passed through the field anode, and the substance is field desorbed in the ionized state. The mass spectrum produced in this way is termed "field desorption mass spectrum" (FD-MS), in contrast to the field ionization mass spectrum (FI-MS). This differentiation of terms is being made with respect to the sample supply. In the case where the sample supply is produced by evaporation from a microcrucible, one speaks of FI-MS, whereas in the case where the supply of the sample is exclusively from the adsorbed layer on the microneedles of the emitter, the term FD-MS is normally used. In principle, with FI a partial supply of the substance occurs by surface diffusion from the shafts of the microneedles to the tips of the needles. However, the differentiation of FI-MS and FD-MS is justified because of the very different techniques of sample supply and because of the different mass spectra observed. In the following the terms FI and FD are used in this sense.

3. *Sensitivity of the FD Method*

The sensitivity of the FD method is defined as the integrated current of the $[M]^{+\cdot} + [M + 1]^{+}$ ions per microgram of the field desorbed sample (C/μg). The maximum sensitivity measured in the laboratories of the authors was about 6×10^{-10} C/μg of a 10^{-7} M solution of cholesterol. The mass resolution was about 500. The mass spectrometer, operated in the double-focusing mode, rendered a sensitivity of about 1×10^{-11} C/μg, which is in qualitative agreement with the value measured by Olson et al. [33] with a Varian MAT 731 mass spectrometer (5×10^{-15} C/ng = 0.5×10^{-11} C/μg) at a mass resolution of 1000.

It must be kept in mind, however, that the absolute ion currents of the $[M]^{+\cdot}$ and/or $[M + 1]^{+}$ ions are rather small in spite of this sensitivity of the FD method. This is because the sample amounts deposited on the field anode are extremely small. On the other hand, the possibility of analyzing these small samples by the FD method is a great advantage for trace analysis. The minimum amount of an organic compound which can be detected with the FD method is about 10^{-11} g at a signal to noise ratio of 3 (if a modified Varian MAT CH4 mass spectrometer is used). This limit depends on the substance, the solvent, its pH value, and the morphological structure of the microneedles as well as on the transmission of the mass spectrometer and the sensitivity of the ion detection system. Therefore, the comparison of the sensitivities for FD of various systems is important.

D. Optimal Field Anode Temperature for FD

Finding the optimal anode temperature T* is very important in FD mass spectrometry. It is not identical to the temperature of the carrier wire because of the temperature gradient between the anode wire and the needle tips. In the case of direct electrical heating of the anode, T* is frequently characterized by describing the heating current (in mA) and called best anode temperature (BAT). Kümmler and Schulten have correlated the heating current with the anode temperature for activated 10 μm tungsten wire of different needle lengths [34] (Fig. 10).

It was found that greater absolute and relative molecular ion currents could be obtained with certain substances by heating the microneedles via thermal radiation [35]. This method of indirect heating of the field anode also has the advantage that the average temperature at the surface of the emitter can be determined within about ±2°C after calibration with negative-temperature-coeffient resistors. The optimal temperature of the field anode, T*, is defined so that the intensity of the molecular ions is maximal and that of the fragment ions is minimal. If a thermally- or field-induced fragmentation of the substance deposited on the field anode is to be carried out, then temperatures will be set above T*, so that the sample

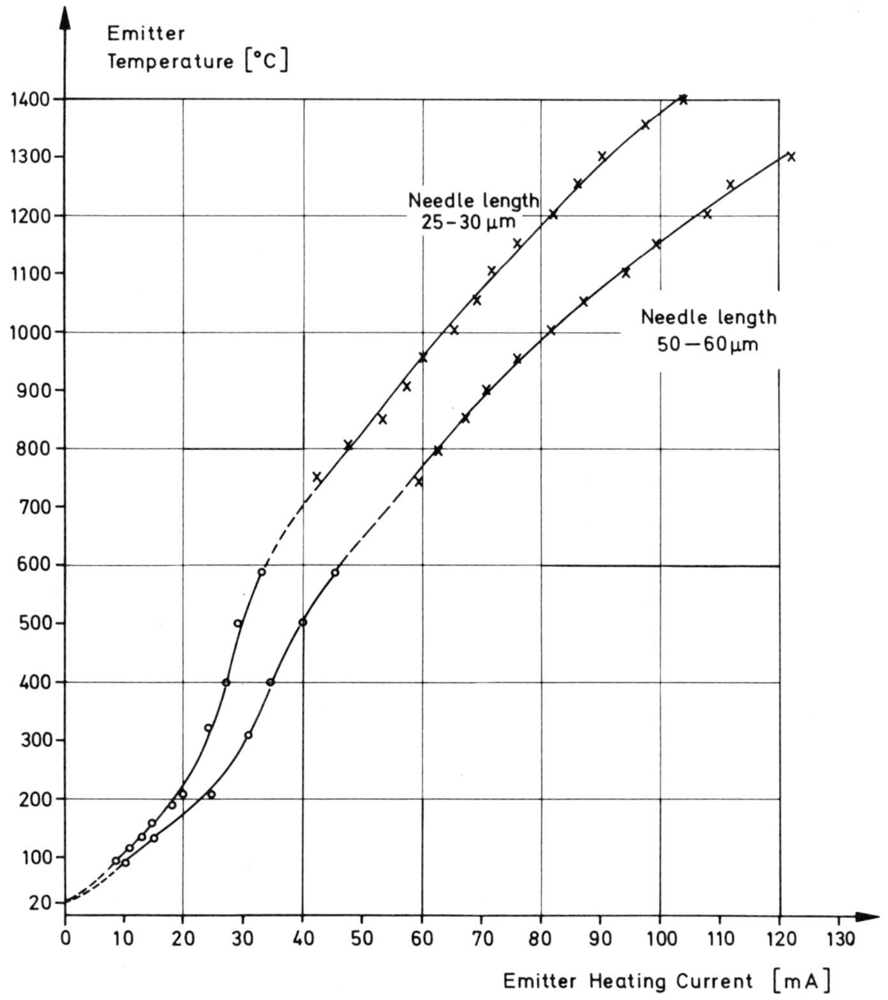

FIG. 10. Graphic plot of the emitter heating current versus the emitter temperature, obtained by melting-point observations in a scanning electron microscope. The dashed parts of the curves were not accessible by the methods employed and have been interpolated [34]. Experiments by Winkler et al. [139] using a different method for the determination of the emitter temperature (with thermocouple and indirect heating) showed that the temperatures given in this figure are probably about 25% too high in the lower-temperature range up to 200°C.

spectrum will be characterized by intense fragment ions. We shall discuss the determination of T* first for the case where the FD spectra are recorded via a slow mass scan with electrical registration and without using the "multiscaling technique." For the molecular weight determination of a sample the optimal anode temperature represents a compromise between two counteracting effects: If the field anode is heated too strongly, then the sample vaporizes too quickly to permit the recording of a complete mass spectrum; however, if the field anode is heated too weakly, then the field desorption rate of the sample is too small for the generation of an ion current which sufficiently exceeds the background noise. T* (expressed by milliamperes of heating current) is dependent on the diameter of the wire, the solvent and its pH value, the thickness of the substance layer on the field anode, and particularly on the length and structure of the microneedles. If the substance is insoluble, it can be suspended in a volatile solvent and deposited on the emitter [36].

T* is determined by first adjusting the mass spectrometer to the mass region of the molecular ion group which is normally dominant in FD-MS. This region is scanned several times and the temperature of the field anode is raised slowly. T* is frequently a very critical quantity, i.e., there is only a very narrow temperature interval dT, where one obtains an intense and relatively easily reproducible mass spectrum. With some substances, T* lies at room temperature, e.g., naphthalene, phenanthrene, and Endrin. With others, T* amounts to several 100°C (i.e., the field anode must be heated with currents of 10-40 mA or even more with certain substances of low volatility). dT is also strongly dependent on the above-mentioned parameters and in particular on the type of substance analyzed. There exist substances, for example, glucosides, where T* is 10 mA and dT is only 0.5 mA. This means that a 5% increase or decrease of T* leads to a field desorption of the sample, which is too fast or too slow.

E. Ion Detection with Photoplates or Fast Electrical Registration

Disadvantages of the FD technique, such as small absolute ion current intensities and unknown optimal temperature for samples of unknown composition, can be overcome with integrating ion detection by means of photoplates or with a fast electrical recorder using the multiscanning technique. Both methods are so well known in EI mass spectrometry, and their advantages and disadvantages have been described so often in the literature, that only the special, additional aspects which appear with the FD technique will be discussed. One method to determine the optimal emitter temperature using photoplates in a double-focusing mass spectrometer is the following: The field anode is heated until an ion current of about 5×10^{-8} A is measured at the counterelectrode (8 kV acceleration voltage, d = 2 mm) [37]. Owing to the simultaneous recording of all ion types on the photoplate, the fluctuations and the general decrease of the ion current as a consequence of the substance consumption affects ions of all m/e values equally. This feature is even more important for the FD technique than for the electron impact method. The fluctuations of the ion current are considerably greater in FD than in EI, and the amount of substance on the field anode is very small (see Sec. III).

Alternatively, an FD mass spectrum can be recorded with the photoplate at temperatures considerably higher than T*. A spectrum is then obtained which shows not only the molecular ion group but also numerous fragment ions [38]. It is possible in this way to draw conclusions as to certain functional groups or branching sites of a molecule of unknown structure. An important field of application for FD mass spectrometry with photoplate ion recording is the pyrolysis of polymers in the range far above T*. Whereas special photoplates (Ilford, Type Q2) were frequently used in the past, gelatine-free photoplates† with a thin layer of silver bromide vaporized onto a glass plate are presently preferable for FD mass spectrometry, particularly in the mass range of $m/e > 500$. With these plates there is considerably less background blackening and hence the signal to noise ratio is much better than with conventional photoplates. Furthermore, the absolute sensitivity of these plates for higher mass ions is greater. Handling the gelatine-free plates is, however, somewhat more critical and their price is considerably higher than that of conventional photoplates.

Instead of ion detection with photoplates, a high-speed electrical recording of ions can be contrived. With the multiscaling technique the mass spectrum is scanned so fast that the intensities of the different types of ions do not change appreciably during one scanning period. Since very few ions per second hit the ion detector during such a fast scan, one has to rely on ion counting techniques, employing, for example, a multichannel analyzer for this purpose. By means of a great number of consecutive scans the various types of ions are stored in the corresponding channels. For expediency, a process computer is used which controls the mass scan through the magnetic field, stores the specific types of ions, and programs the temperature of the field anode. In the authors' laboratories a process computer, type PDP 8/E, Digital Equipment Co., is used.

---

†Manufactured by Ionomet Co., P. O. Box 56, Waban, MA.

F. Means of Distinguishing Molecular Ions, $[M]^{+\cdot}$, and Protonated Molecules, $[M + H]^+$

A certain difficulty in the determination of molecular weight by FD-MS arises from the fact that apart from the molecular ions, $[M]^{+\cdot}$, very intense $[M + H]^+$ ions are often observed. They are formed, especially with highly polar compounds, by proton transfer reactions at the surface of the field anodes [16]. If one is sure by certain prior information that a sample does not contain nitrogen, then this difficulty does not appear, because the $[M]^{+\cdot}$ ion must have an even mass, and the $[M + H]^+$ ion an odd mass (i.e., the nitrogen rule). If the presence of nitrogen atoms in the molecule cannot be excluded, the situation becomes more complicated. Generally, the decision as to whether the peak in question is an $[M]^{+\cdot}$ or an $[M + H]^+$ peak can be reached through the variation of several parameters, such as field strength, emitter temperature, pH value of the solvent. In addition, the decision can be aided through the extraction of other information from the mass spectra, such as measurement of metastable ions, analysis of the shape of mass signals, detection of doubly charged ions, $[M - H]^+$ ions, cluster ions, $[nM + H]^+$, as well as "cationization" of molecules by traces of inorganic salts [39], and finally through knowledge about the presence of nucleophilic groups in the molecule. As many as possible of the listed criteria must be consulted for the differentiation between the $[M]^{+\cdot}$ and the $[M + H]^+$. The details of this problem cannot be discussed within the framework of this review article. Reference is therefore made to an original contribution on this subject [40].

V. ANALYTICAL APPLICATIONS OF FIELD IONIZATION MASS SPECTROMETRY

A comprehensive account of the analytical applications of FI mass spectrometry up to 1969 was given in a review article [41] and in the monograph by Beckey [1]. Therefore, only results which have been published after 1969 will be described in the following sections.

A. FI-MS of Model Compounds

Several authors have studied model compounds in order to elucidate the applicability of the FI method for structure analysis of unknown compounds. It was pointed out that the FI mass spectra are complementary to the EI mass spectra in those cases where the molecular ion cannot be observed or is ambiguous in the EI spectra.

Krone and Beckey [42] have studied mono- and oligosaccharides with FI in order to develop the use of the FI method for structure analysis and for elucidation of the mechanisms of the ionized molecules. Molecular ions of saccharides, or their derivatives, are not (or with small intensities) detectable in the EI spectra [43].

1. Sugars

The FI mass spectra of the unsubstituted D-glucose, D-galactose, D-fructose, and D-xylose exhibit a very intense $[M + 1]^+$ peak which is also the base peak if the temperatures of evaporation of these compounds are not too high. The relative intensity of the molecular ions is only a few percent in some cases. $[M + 1]^+$ and $[M - 1]^+$ ions are formed by proton transfer processes at the surface of the field anode. The strength of an O-H bond is decreased so much by FI that proton transfer to a proton acceptor which is nearby can occur. One can conclude from studies of deuterated alcohols that the proton is preferrentially split off from a hydroxy group. The [M - 1] radical can be field ionized and also chemisorbed on the surface of the emitter causing secondary reactions. Therefore, the intensity of the $[M - 1]^+$ ions is much smaller than that of the $[M + 1]^+$

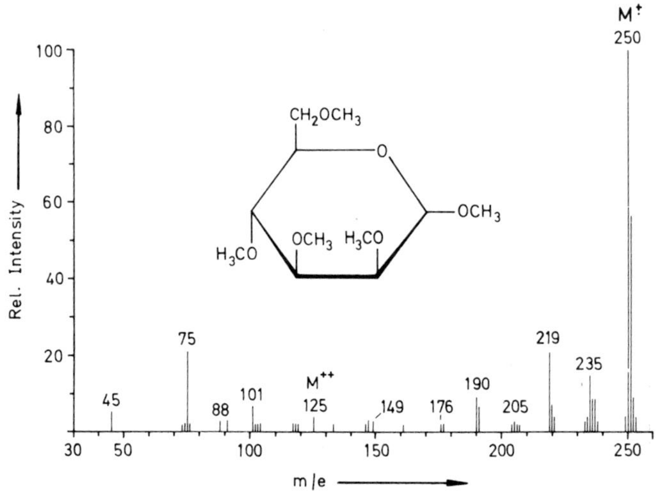

FIG. 11. FI mass spectrum of methyl-2,3,4,6-tetra-O-methyl-D-mannopyranoside [42].

ions. Heyns and Müller [44] have shown that the majority of the fragment ions in the case of EI are formed by consecutive reactions which are initiated by rupture of the C(1)-C(2) bond. These reactions are rearrangement reactions with few exceptions. The FI spectrum of methyl-2,3,4,6-tetra-O-methyl-D-mannopyranoside is shown in Fig. 11. Here, in contrast to the free monosaccharides, the molecular ion peak is the base peak of the spectrum. The proton transfer reaction is made impossible by the introduction of methoxy groups. $[M + 1]^+$ ions can be formed with permethylated monosaccharides only by reactions where a proton is split from an α-C-H bond. Unsubstituted disaccharides are not well suited for taking FI mass spectra because of their thermal instability. In the FI spectrum of hepta-O-methyl-primverose, however, the molecular ion peak $[M]^{+\cdot}$ is the base peak. FI spectra with a relatively intense molecular ion peak can also be taken with trisaccharides if they are permethylated. Generally, however, obtaining an FD spectrum is more favorable with oligosaccharides. The FD spectrum of cellobiose, for example, exhibits the $[M + 1]^+$ peak as base peak [42].

2. Other Biologically Significant Compounds

Damico, Barron, and Sphon [45] carried out a very profound comparative study with FI and EI on some *pesticidal* and other *biologically significant compounds*. They confirmed the general finding that FI and EI spectra are complementary to each other. The molecular ion (or the $[M + 1]^+$ ion) was found with high intensity in those cases where no such ions were found unambiguously with EI. The FI mass spectra of chlorinated pesticides like Endrin (Fig. 12) are especially interesting because they can be compared with the FD spectra measured by Schulten and Beckey [46].

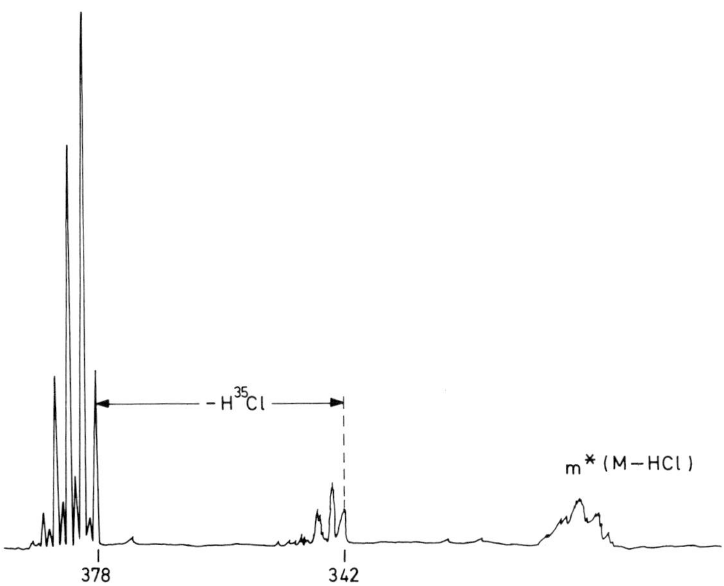

FIG. 12. Selected portion of the electrically recorded FI mass spectrum of Endrin; sample temperature 20°C [45].

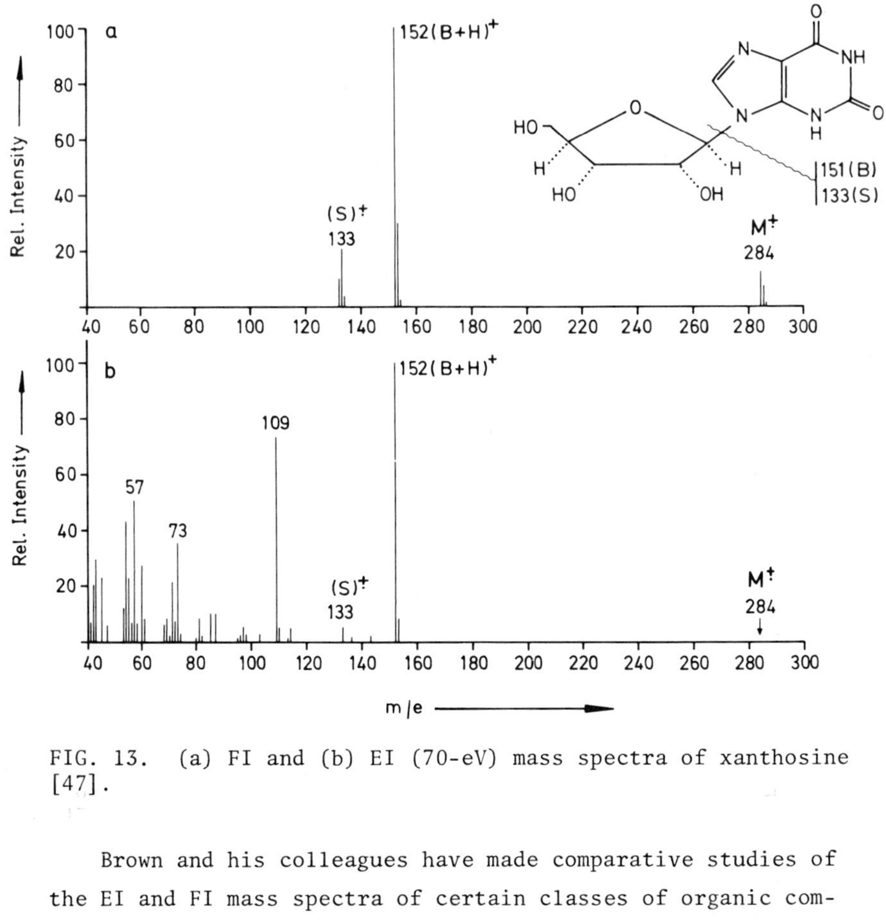

FIG. 13. (a) FI and (b) EI (70-eV) mass spectra of xanthosine [47].

Brown and his colleagues have made comparative studies of the EI and FI mass spectra of certain classes of organic compounds. Brown, Pettit, and Robins compared the EI and FI spectra of a number of *nucleosides* [47]. They found that the EI spectra are complicated and that the molecular ions and sugar fragments which are important for structure analysis frequently have low intensities or are undectable. In contrast, the FI spectra are extremely simple and intense molecular ion peaks are commonly observed. This point is clearly demonstrated in the FI (Fig. 13a) and EI (Fig. 13b) mass spectra of xanthosine. However, an exception is guanosine. Here the FI spectrum shows no molecular ion at all, whereas with FD-MS about 20% relative intensity of this ion is observed. Furthermore, Brown and Pettit [48] have studied the benzyloxycarbonyl and t-butyloxycarbonyl derivatives of some simple *peptides*. The derivatives of di- to pentapeptide methylesters have been studied which contain the amino acids glycine, alanine, leucine, serine, threonine, proline, and tyrosine. The sequence characteristic fragment and rearrangement peaks were somewhat less intense in the FI than in the EI spectra, whereas the intensities of the molecular ion peaks were higher in the FI spectra.

The most important and extensive FI work of Brown et al. [49, 50] was concerned with *cardenolides*. The EI and FI spectra of cardenolides of the cardiac glycoside type were studied. These physiologically active compounds contain a steroidal genine (G) (or aglycone) attached by a 3-O-glycoside linkage to a mono- to pentasaccharide moiety. The mass spectra of underivatized compounds were of primary interest. The major objective was characterization of the genine moiety and of the component monosaccharides and their sequence. As an example, only the FI (Fig. 14a) and EI (Fig. 14b) mass spectra of digitoxine will be discussed here. The genine is digitoxigenine in this case. The side chain consists of three d-digitoxose moieties. The molecular weight of this compound is relatively high (M = 764). The molecular peak cannot be detected unambiguously because its relative intensity is only about 10^{-4} of the base peak in the EI spectrum; in the FI spectrum, however, its relative intensity is about 3%. Moreover, the characteristic sugar sequence peaks exhibit a larger relative intensity in the FI spectra. Generally, the molecular weights and the sugar sequence peaks of up to three underivatized sugar moieties can be determined if the total number of free hydroxy groups is not too large in the cardiac glycoside. Otherwise derivatization is recommended for such polar material.

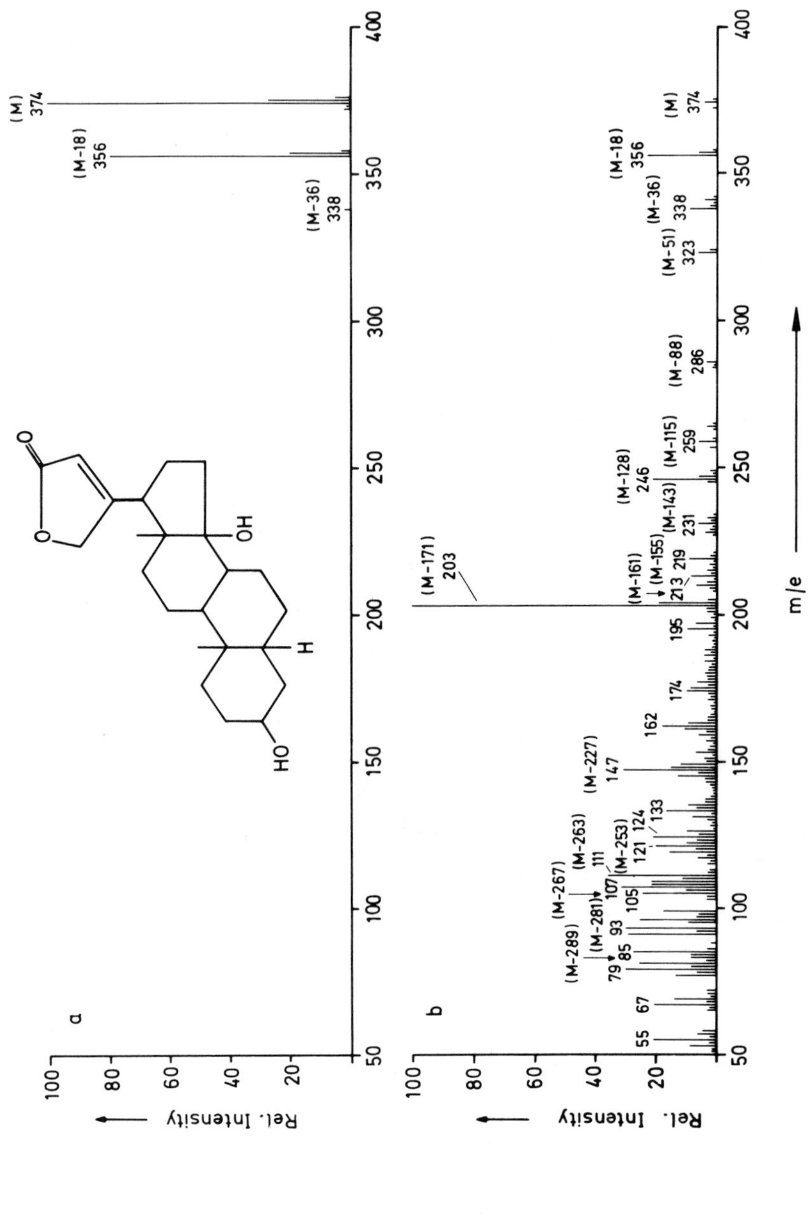

FIG. 14. (a) FI and (b) EI (70-eV) mass spectra of digitoxine [50].

Fenselau, Wang, and Brown [51] have compared the EI and FI mass spectra of *photopolymers of thymine*. They observed a cis-syndimeric ion of thymine, a thymine-thymine adduct, and a dimeric ion of thymine. In this case the FI spectra were more characteristic than the EI spectra, particularly with respect to the selectivity of the fragment ions. Brown and Fenselau [52] further studied the FI mass spectra of isotopically labeled heptanals. This work was more concerned with the kinetic and mechanistic aspect of decomposition of field ionized molecules. It was found, for example, that the molecular ions have not undergone significant H/D randomization at extremely short decomposition times.

Chait et al. [53], in contrast to the authors cited so far, have used activated razor blades as field anodes for analysis of a variety of natural compounds. They obtained a resolving power of $m/\Delta m = 20{,}000$ with a double-focusing mass spectrometer.

It should be mentioned in this review that Goldenfeld et al. [54] have studied the FI mass spectra of a number of classes of organic compounds, such as nitrogen-containing compounds, fluoroparaffins, aldehydes, epoxy compounds, amines, ethers, and nitriles. Interestingly, these authors found that the molecular ions of $[CF_4]^{+\cdot}$ and $[C_2F_6]^{+\cdot}$ are completely unstable, so that they cannot be detected in the FI mass spectra even at the highest sensitivity of the FI mass spectrometer (dynamic range $1{:}10^5$). If only one of the fluorine atoms is replaced, however, by a hydrogen, chlorine, or iodine atom, one can observe molecular ions. The finding of Goldenfeld et al. on the other classes of organic compounds agree generally with those of the previously mentioned authors.

There have been more recent publications in the area of the comparison of various ionization techniques by Fales et al. [55] and Haskins et al. [56]. These authors compared the EI, CI, FI, and FD mass spectra of some biologically significant compounds.

The advantages and disadvantages of the various methods of ionization were discussed. It was pointed out that no technique of ionization is uniformly superior to the others. The method of choice in any analytical problem will depend on the type of compound involved. Very often the methods prove to be complementary.

B. FI-MS of Mixtures

In the discussion of the principle of FI (Sec. II) and in the studies of the model compounds by FI-MS noted above there is one common feature: If the substance can be transferred into the gas phase without decomposition, then, owing to the small electronic and vibrational excitation, the mass spectra obtained are generally characterized by the presence of prominent molecular ion peaks and of only a few minor fragment ion peaks.

FI-MS is therefore, in principle, a suitable technique for the analysis of multicomponent mixtures of volatile organic compounds. This was demonstrated by Beckey and Wagner in 1963 [57]. In a small comparative study of the qualitative and quantitative analysis of a seven-component hydrocarbon mixture by gas-liquid chromatography (GLC) and FI-MS. That this combination is a particularly useful tool for the analysis of mixtures was confirmed by the work of Damico and Barron [58]. These authors used a GLC FI-MS system incorporating a permeable membrane interface to obtain spectra of GLC effluents consisting of components that yield relatively weak molecular ions by EI-MS. FI sensitivity studies of the molecular ion for a variety of carbonyl compounds suggest that some types of compounds are significantly more sensitive to field ionization than others.

For a sensitivity test carbonyl compounds were studied that occur in small amounts in food extracts. The results shown in Table 1 clearly indicate significant differences in the FI sensitivity for various types of carbonyl compounds. Thus, for

TABLE 1

Field Ionization Sensitivity for the Molecular Ion of Some Carbonyl Compounds

Compound	Amount injected (μg)	Maximum molecular ion response (V/μg)	Relative intensity[a] of molecular ion (%) FI	EI
C_8-Δ-Lactone	50	45	100	6.6
C_{10}-γ-Lactone	30	100	100	0.3
C_{10}-Δ-Lactone	30	250	100	3.5
C_{12}-γ-Lactone	30	88	100	0.1[b]
C_{12}-Δ-Lactone	30	110	100	4.3
C_{14}-Δ-Lactone	50	200	100	<0.1[b]
C_{10} Acid	250	12	100	4.0[c]
Decylacetate	500	0.2	100	<0.3
C_{10}-Methylester	40	62	100	3.6[b]
C_{12}-Methylester	40	62	100	2.7
C_{14}-Methylester	40	62	100	3.2
C_9-Ketone	50	70	100	4.9
C_{11}-Ketone	50	200	100	2.5[b]
C_{15}-Ketone	50	200	100	3.0[d]
C_8-Aldehyde	50	0.4	100	1.0
C_{10}-Aldehyde	50	0.8	100	0.5
C_{12}-Aldehyde	50	2	100	0.3

[a]For references see ASTM uncertified Mass Spectral Data File [58].
[b]Obtained with electron impact E-4B source using GLC-MS inlet.
[c]Obtained with electron impact E-4B source using direct probe.
[d]Obtained with EFO-4B source using GLC-MS inlet.

quantitative results the calibration of the GLC FI-MS system may give considerable problems.

Another approach for using FI-MS for mixture analysis is the direct introduction of the mixture into the ion source of the FI mass spectrometer avoiding possible discrimination by the GC column and separator. Complex mixtures were studied by Schüddemage and Hummel [59]. These authors analyzed pyrolysates of synthetic polymers by FI-MS and their results illustrate the striking simplicity of FI-MS spectra of multicomponent mixtures as compared to the corresponding EI mass spectra.

It should be pointed out, however, that the interpretation of spectra obtained by low-resolution FI-MS of multicomponent mixtures is inevitably complicated by the fact that molecular ions of different elemental composition may share the same nominal m/e value. High-resolution FI-MS therefore greatly widens the scope of the method since it permits the determination of the elemental composition of the observed ions by accurate mass measurements. Forehand and Kuhn [60] performed high-resolution FI-MS of the condensable phase of cigarette smoke and proposed probable elemental compositions for more than 50 components.

Direct analysis by FI-MS with high resolution and photographic recording of the products generated in the pyrolysis of bacteria gave spectra that contained the molecular ions of more than 200 individual chemical species [61] (Fig. 15). As may be inferred from Table 2, among other compounds free acids, amines, alcohols, and branched hydrocarbons were identified. These results show that the range of compounds that can be analyzed by direct FI-MS of multicomponent mixtures is much broader than that attained by any single GLC-(EI)MS combination. In practice, however, the usefulness of high-resolution FI-MS analysis is limited by the inability to separate and identify isomers, unless additional information is obtained from metastable processes, fragment ions, or doubly charged ions. In this respect a combination of EI-MS and FI-MS with gas chromatography has been successfully applied to crude mixtures of alkaloids [62] and coumarins [63] isolated from natural sources.

Recently the combined investigation of pyrolysis products from RNA and DNA by low-voltage-ionization quadrupole mass spectrometry and high-resolution FI-MS were reported [64]. The spectra show that most of the pyrolysis products originate from the carbohydrate moiety of the nucleic acids. Micrograms of RNA/DNA can be distinguished very easily by the use of key fragments, and reaction mechanisms for the pyrolysis of the polymers which

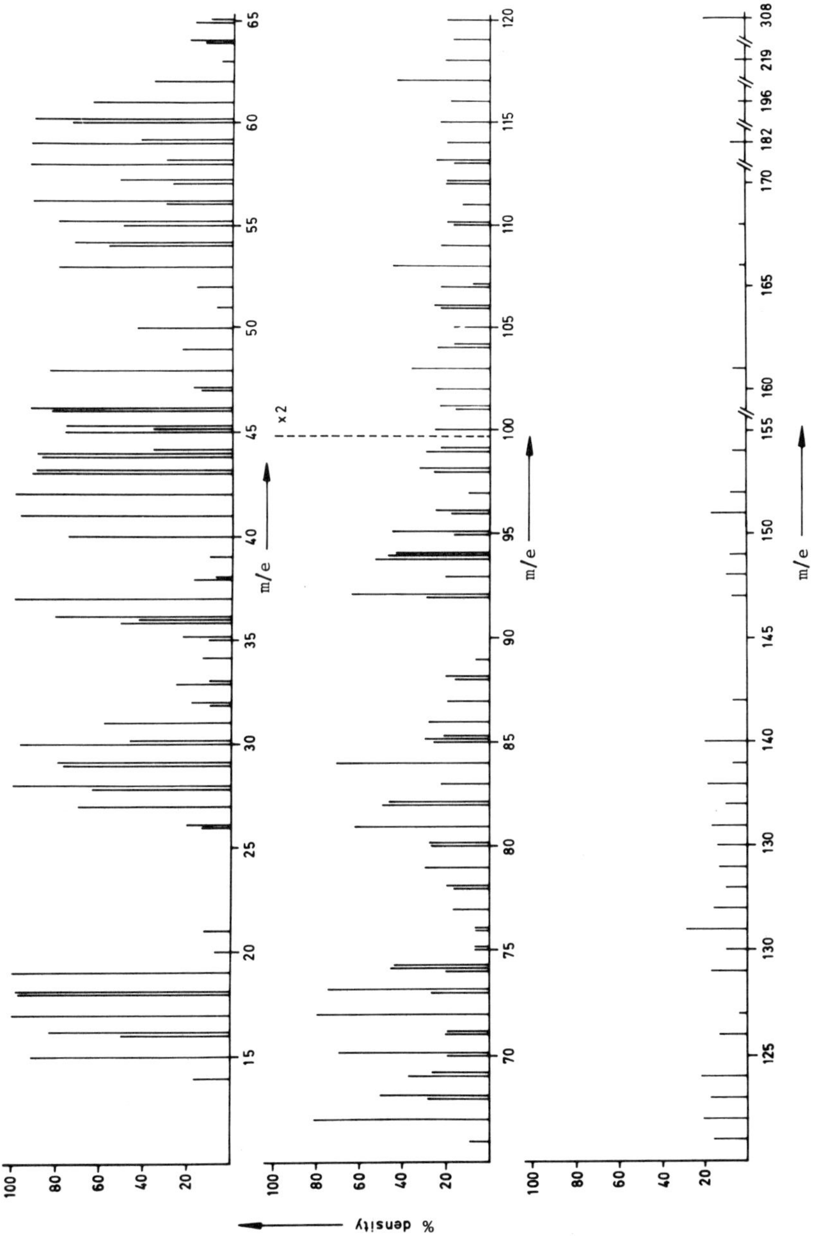

FIG. 15. High-resolution FI mass spectrum of pyrolysis products of *Pseudomonas putida* bacteria [61]; the mass distances are not drawn to scale, 100% density corresponds with the saturation blackening value of the photoplate (in this case Ilford Q_2 plates were used).

TABLE 2

Pyrolysis Products of *Pseudomonas* Bacteria[a]

Measured mass	Probable elemental composition	Ion type[b]	Proposed compounds
14.015	CH_2	f	
15.024	CH_3	f	
16.014	NH_2	f	
16.030	CH_4	M	Methane[c]
17.027	NH_3	M	Ammonia[c]
18.011	H_2O	M	Water[c]
18.035	$NH_3 + H^+$	M + 1	
19.018	$H_2O + H^+$	M + 1	
20.013	$H_2{}^{18}O$	i	
21.020	$H_2{}^{18}O + H^+$	i	
26.002	CN	f	
26.015	C_2H_2	M	Ethyne
27.010	HCN	M	Hydrocyanic acid
27.994	CO	M	Carbon monoxide
28.028	C_2H_4	M	Ethene[c]
29.005	CHO	f	
29.040	C_2H_5	f	
30.005	CH_2O	M	Formaldehyde
30.048	C_2H_6	M	Ethane
31.038	CH_3NH_2	M	Aminomethane
31.991	O_2	M	Oxygen
32.026	CH_3OH	M	Methanol[c]
32.980	HS	f	
33.021	NH_2OH	M	Hydroxylamine
33.986	H_2S	M	Hydrogen sulfide[c]
35.036	NH_4OH	M	Ammonium hydroxide
35.060	$2NH_3 + H^+$	c	
35.976	$H^{35}Cl$	M	Hydrochloric acid
36.018	$2H_2O$	c	
36.042	$NH_4OH + H^+$	M + 1	
37.028	$2H_2O + H^+$	c	
37.974	$H^{37}Cl$	i	
38.016	C_3H_2	f	
39.022	C_3H_3 or $C_6H_6{}^{2+}$	f or M^{2+}	
40.020	C_2H_2N	f	
41.027	C_2H_3N	M	Ethanenitrile[c]
42.044	C_3H_6	M	Propene[c]
43.016	C_2H_3O	f	
43.060	C_3H_7	f	
43.991	CO_2	M	Carbon dioxide[c]
44.022	C_2H_4O	M	Ethylene oxice[c]

TABLE 2 (continued)

Measured mass	Probable elemental composition	Ion type[b]	Proposed compounds
44.064	C_3H_8	M	Propane
45.003	CHO_2	f	
45.020	CH_3NO	M	Formamide
45.029	C_2H_5O	M + 1	
46.008	CH_2O_2	M	Formic acid
46.035	C_2H_6O	M	Ethanol
47.002	CH_3S	f	
47.030	CH_5NO	M	Hydroxyamino-methane
48.004	CH_4S	M	Methanethiol[c]
49.015	$CH_4S + H^+$	M + 1	
49.997	$CH_4{}^{34}S$	i	
51.006	$CH_4{}^{34}S + H^+$	i	
52.002	C_2N_2	M	Cyanogen
53.025	C_3H_3N	M	Propenenitrile[c]
54.011	C_3H_2O	M	Propynal
54.047	C_4H_6	M	Butadiene[c]
55.020	C_3H_3O	f	
55.043	C_3H_5N	M	Propanenitrile[c]
56.024	C_3H_4O	M	Propenal[c]
56.060	C_4H_8	M	Butene[c]/Methyl-propene[c]
57.032	C_3H_5O	f	
57.068	C_4H_9	f	
58.043	C_3H_6O	M	Acetone[c]/Propanal[c]
58.076	C_4H_{10}	M	Butane[c]
59.040	C_2H_5NO	M	Acetamide[c]
59.052	C_3H_7O	M + 1	
60.022	$C_2H_4O_2$	M	Acetic acid
60.045	$C_2H_5NO + H^+$	M + 1	
61.03]	$C_2H_4O_2 + H^+$	M + 1	
62.023	C_2H_6S	M	Ethanethiol/Di-methylsulfide
63.003	HNO_2	M	Nitric acid
63.960	SO_2	M	Sulfur dioxide
64.009	$HNO_2 + H^+$	M + 1	
64.968	$SO_2 + H^+$	M + 1	
65.010	$HNO_3 + 2H$	M + 2	
66.030	C_4H_4N	f	
67.039	C_4H_5N	M	Pyrrole[c]/Methyl-propenenitrile[c]
68.023	C_4H_4O	M	Furan[c]
68.059	C_5H_8	M	Methylbutadiene[c]

TABLE 2 (continued)

Measured mass	Probable elemental composition	Ion type[b]	Proposed compounds
69.013	C_3H_3NO	M	Hydroxypropenenitrile
69.057	C_4H_7N	M	Butanenitrile[c]/ Methylpropanenitrile[c]
70.008	$C_3H_2O_2$	M	Propynoic acid
70.075	C_5H_{10}	M	Methylbutene[c]/ Pentene[c]
71.014	$C_3H_2O_2 + H^+$	M + 1	
71.040	C_3H_5NO	M	Hydroxypropanenitrile
72.058	C_4H_8O	M	Butanone[c]/Methylpropanal[c]
73.007	C_3H_5S	f	
73.057	C_3H_7NO	M	Propionamide[c]
74.017	C_3H_6S	M	Thiapropane[c]/Propenethiol/Vinylmethylsulfide
74.042	$C_3H_6O_2$	M	Propionic acid/ Hydroxypropanal
74.067	$C_3H_7NO + M^+$	M + 1	
75.004	$C_3H_5{}^{34}S$	i	
75.051	$C_3H_6O_2 + H^+$	M + 1	
76.007	$C_3H_6{}^{34}S$	i	
76.031	C_3H_8S	M	Propanethiol/Ethylmethylsulfide
77.069	$C_2H_8N_2O + H^+$	M + 1	
78.015	C_2H_6OS	M	Hydroxyethanethiol
78.047	C_6H_6	M	Benzene[c]
79.043	C_5H_5N	M	Pyridine[c]
80.042	$C_4H_4N_2$	M	Pyrazine[c]
80.052	$C_5H_5N + H^+$	M + 1	
81.060	C_5H_7N	M	Methylpyrroles[c]
82.043	C_5H_6O	M	Methylfuran[c]
82.065	$C_5H_7N + H^+$	M + 1	
83.073	C_5H_9N	M	Methylbutanenitrile[c]
84.091	C_6H_{12}	M	Methylpentene[c]/ Hexene[c]
85.050	C_4H_7NO	M	Hydroxybutanenitrile
85.063	C_5H_9O	f	
85.076	$C_4H_8N_2 + H^+$	M + 1	
86.071	$C_5H_{10}O$	M	Pentanone[c]/Methylbutanal[c]

TABLE 2 (continued)

Measured mass	Probable elemental composition	Ion type[b]	Proposed compounds
87.075	C_4H_9NO	M	Butyramide
88.051	$C_4H_8O_2$	M	Butanoic acid/Hydroxybutanal
88.074	$C_4H_9NO + H^+$	M + 1	
89.080	$C_4H_{11}NO$	M	Aminobutanal
92.045	$C_3H_8O_3$	M	Glycerol
92.060	C_7H_8	M	Toluene[c]
93.058	C_6H_7N	M	Methylpyridine[c]
93.986	$C_2H_6S_2$	M	Dimethyldisulfide[c]
94.036	C_6H_6O	M	Phenol[c]
94.056	$C_5H_6N_2$	M	Methylpyrazine
95.034	C_5H_6NO	M	Hydroxypyridine
95.074	C_6H_9N	M	Dimethylpyrroles[c]
96.026	$C_5H_4O_2$	M	Furfural[c]
96.064	C_6H_8O	M	Dimethylfuran[c]
97.057	C_5H_7NO	M	Furfurylamine
98.038	$C_5H_6O_2$	M	Furfurylalcohol[c]
98.114	C_7H_{14}	M	Heptene[c]
99.042	$C_5H_6O_2 + H^+$	M + 1	
99.077	$C_6H_{11}O$	f	
100.085	$C_6H_{12}O$	M	Methylpentanone[c]
101.078	$C_5H_{11}NO$	M	
101.085	$C_6H_{12}O + H^+$	M + 1	
102.089	$C_5H_{11}NO + H^+$	M + 1	
103.041	C_7H_5N	M	Benzenenitrile[c]
104.046	$C_7H_5N + H^+$	M + 1	
104.062	C_8H_8	M	Styrene[c]
105.052	C_7H_7N	M	
106.060	$C_7H_7N + H^+$	M + 1	
106.073	C_8H_{10}	M	Xylenes[c]/Ethylbenzene[c]
107.069	C_7H_9N	M	Dimethylpyridine[c]
107.084	$C_7{}^{13}CH_{10}$	i	
108.058	C_7H_8O	M	Cresoles[c]
109.090	$C_7H_{11}N$	M	C_3-alkylpyrroles[c]
110.078	$C_7H_{10}O$	M	
110.096	$C_7H_{11}N + H^+$	M + 1	
111.062	C_6H_9NO	M	
112.081	$C_7H_{12}O$	M	
112.125	C_8H_{16}	M	
113.015	$C_4H_3NO_3$	M	
113.090	$C_5H_{11}NO$	M	
114.100	$C_7H_{14}O$	M	

TABLE 2 (continued)

Measured mass	Probable elemental composition	Ion type[b]	Proposed compounds
115.058	$C_5H_9NO_2$	M	
116.063	C_9H_8	M	Indene[c]
117.068	C_8H_7N	M	Indole[c]/Tolunitrile[c]/ Phenylacetonitrile[c]
118.078	C_9H_{10}	M	Methylstyrene[c]
119.083	C_9H_{10} + H^+	M + 1	
120.092	C_9H_{12}	M	C_3-alkylbenzenes[c]
121.101	C_9H_{12} + H^+	M + 1	
122.076	$C_8H_{10}O$	M	Ethylphenol[c]/ Xylenoles[c]
123.102	$C_8H_{13}N$	M	C_4-alkylpyrroles[c]
124.109	$C_8H_{13}N$ + H^+	M + 1	
126.130	C_9H_{18}	M	C_9-alkenes[c]
127.142	$C_8{}^{13}CH_{18}$	i	
129.052	C_9H_7N	M	
130.092	$C_7H_{14}O_2$	M	
131.070	C_9H_9N	M	Methylindole[c]
132.072	$C_8{}^{13}CH_6N$	i	
133.092	$C_9H_{11}N$	M	
134.077	$C_8H_{10}N_2$	M	
135.048	$C_4H_9NO_4$	M	
136.090	$C_9H_{12}O$	M	
137.084	$C_8{}^{13}CH_{12}O$	i	
138.071	$C_8H_{10}O_2$	M	
139.083	$C_8H_{10}O_2$ + H^+	M + 1	
140.109	$C_9H_{16}O$	M	
142.093	$C_8H_{14}O_2$	M	

[a]High-resolution FI-MS data and proposed chemical identity of the observed molecular ions [61].
[b]M = Molecular ion. M + 1 = protonated molecule. f = fragment ion. i = isotope-containing ion. c = complex ion.
[c]Pyrolysis products identified by Simmonds in Py-GLC-MS studies of bacteria and organic soil material (for references see Ref. 61).

may shed some light on pyrolysis processes. As shown in Scheme 1 the formation of m/e 82 from DNA is explained by the interaction between the base and the 5'-phosphate ester as the initial step. The expulsion of the 3'-phosphoric acid ester through

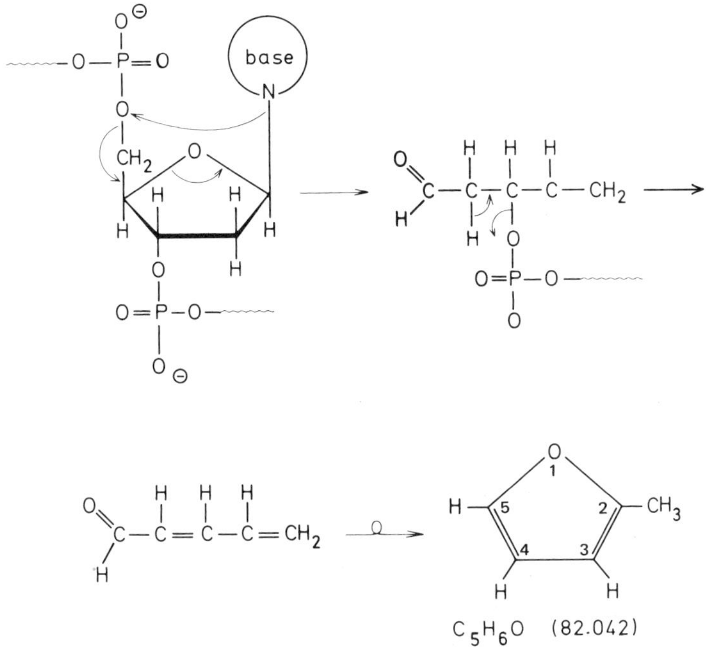

SCHEME 1. Rationalization of the thermal formation of mass 82 (C_5H_6O) from DNA [64].

1,2 elimination is suggested as a possible mechanism. These thermal degradations finally generate 2-methylfuran. Unequivocal evidence for this final product is given by collisional activation spectra [65].

In view of the considerable capacity of FI-MS for analyzing single compounds and mixtures that can be evaporated without decomposition, it is astonishing that the use of the FI method for analytical purposes has been restricted to a relatively small number of research groups. Two main facts may have contributed to this: The first crucial point is the difficulty in preparing efficient and mechanically as well as chemically stable emitters. Second, in the early times of FI-MS the limited access to commercially available instruments made it necessary to work with homemade FI instrumentation providing the money and know-how were available. Because the preparation of high-temperature-activated emitters has become a routine procedure and commercial dual EI/FI-FD sources are available, a widespread analytical application of FI-MS might be expected. This holds particularly for the coupling of a GLC system and a high-resolution mass spectrometer equipped with photographic detection. For compounds that cannot be volatilized, FD-MS appears to be the appropriate technique as will be discussed in Sec. VI.

VI. APPLICATION OF FIELD DESORPTION MASS SPECTROMETRY

The development of mass spectrometry during the last 10 years shows an increasing tendency toward applications in the biochemical field. Emphasis is being placed particularly on the analytical capacities of new techniques and methods in the areas of environmental research, biochemistry, and medicine. Especially strict requirements must be met in these investigations.

1. The *detection sensitivity* of a technique should be in or below the nanogram range. This is of particular importance for the chemical-analytical problems of metabolization and for the trace analysis of environmental chemicals.
2. The *selectivity* of the method should be such that even substances in only prepurified crude extracts of biological material can be detected unequivocally along with the unavoidable admixtures.
3. With respect to the *significance* of the MS data, it is required that, apart from characteristic fragment ions, molecular ion intensities as high as possible be produced, since in the above-described areas, in many cases, the determination of the molecular weight of an unknown compound, or rather its elemental composition, is the main aim of the investigation.
4. *Reliability* of the procedure in qualitative and quantitative analysis is required. This is rendered especially difficult, since products from the physiological environment have

generally little thermal stability (besides being sensitive to oxidizers, changes in the pH value, etc.). Moreover, a great number of free functional groups are usually present and are coupled with strong intermolecular forces, which lead to low volatility. In many cases it is therefore very difficult or even impossible to transfer the intact molecule undecomposed into the gas phase. This, however, is the basic requirement of conventional mass spectrometric methods such as EI, CI, and FI. In order to demonstrate the contributions of the unique features of FD-MS and its specific abilities as an *additional* tool, we shall treat by way of example three areas in which the successful application of the FD technique becomes apparent.

A. Application in Environmental Analysis

1. *Pesticides*

In recent years the mass spectrometer has proved itself successful in the qualitative and quantitative detection of toxic substances in the environment. Its high and uniformily good selectivity in the analysis of mixtures of samples have been shown to be highly advantageous. This development has been even more enhanced by the possibilities of coupling the MS to a gas chromatograph and to a fast electronic data recording and data processing computer. Chlorinated pesticides, organic phosphorous pesticides, polychlorinated biphenyls, and organic mercury compounds were the main aim of the EI-MS investigations in this area. However, it was seen that, with a considerable number of the investigated underivatized compounds, the molecular ions could hardly be detected or did not appear at all in the mass spectrum. Since field ionization had matured in the meantime into a well-operating ionization method [41] it was possible to demonstrate the capacities of the new method in the analysis of chemicals occurring in the environment. Damico et al. [45] were able to demonstrate in 1969 in a first comparison of EI and FI mass spectra of pesticides that much higher relative intensity molecular ions could be detected using the FI technique. Since the samples have to be vaporized with EI-MS as well as with FI-MS, the decisive difference between the two ionization methods is found in the considerably smaller energy transferred to the molecule in the ionization process (∿5 eV EI-MS and a few tenths of an electron volt with FI-MS). Owing to this small transferred average internal energy, the molecular ions formed with FI produce, in general, only a few fragments. These correspond very often to the most important EI fragments and yield valuable structural information. It became quickly obvious, however, that even with the FI technique a number of important pesticides showed only relatively small molecular ion intensities. This happens when the thermal energy, which is needed for vaporizing the solid sample, causes thermal decomposition *prior* to ionization. Therefore pyrolysis products reach the ionization chamber where they are ionized and appear in the mass spectrum. It then becomes very difficult or impossible to draw conclusions as to the structure of the actual compound investigated, particularly in contaminated samples.

a. Chlorinated polycyclic hydrocarbons. Figure 16 shows the EI-MS of the insecticide, acaricide, and rodenticide Endrin[†] [66]. The molecular ion at m/e 378 is registered with about 1% relative intensity. In the FI mass spectra of a series of chlorinated polycyclic hydrocarbons recorded by Damico et al. [58,66] it was seen that the molecular ions of these compounds make up the prevailing share of the produced ion current. In the case of Endrin the volatility of the substance is sufficient to obtain an FI mass spectrum at 20°C. Apart from the intensive

---

[†]The common names have been used for the cited pesticides. The exact chemical nomenclature is to be found in O. R. Klimmer, *Pflanzenschutz- und Schädlingsbekämpfungsmittel, Abriss einer Toxikologie und Therapie von Vergiftungen*, Hundt-Verlag, Hattingen, 1971 (in German), and in S. Safe, O. Hutzinger, *Mass Spectrometry of Pesticides and Pollutants*, CRC Press, Cleveland, Ohio, 1973 (in English).

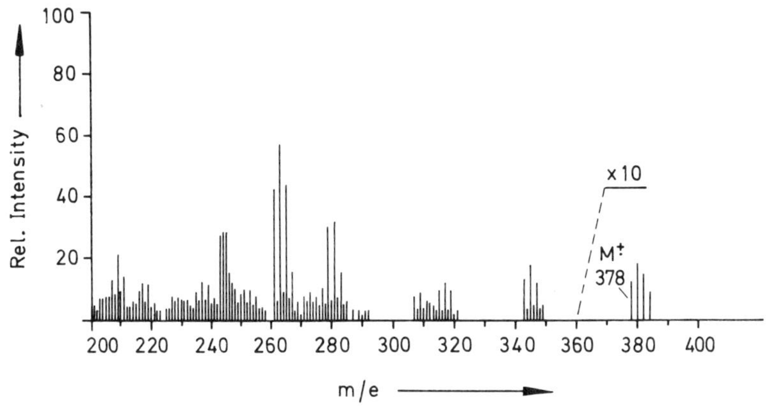

FIG. 16. EI mass spectrum of Endrin [66].

molecular ion group (base peaks), the signal at m/e 342 is observed with about 10% relative intensity. From the corresponding metastable peak the indication for HCl elimination is obtained. For comparison it should be mentioned that the four most important signals in the CI-MS of Endrin are noted by Biros et al. [67] at m/e 397 (9% rel. int., $[M + H_3O]^+$), m/e 379 (19%

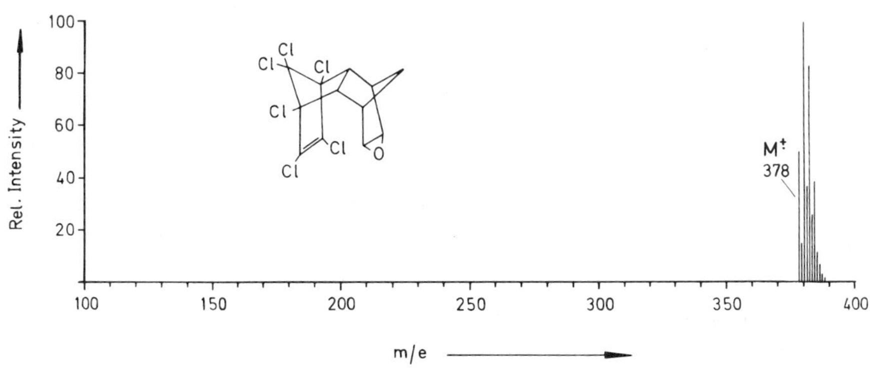

FIG. 17. FD mass spectrum of Endrin. Solvent acetone, recorded with a low-resolution mass spectrometer with electric detection (at room temperature) [68].

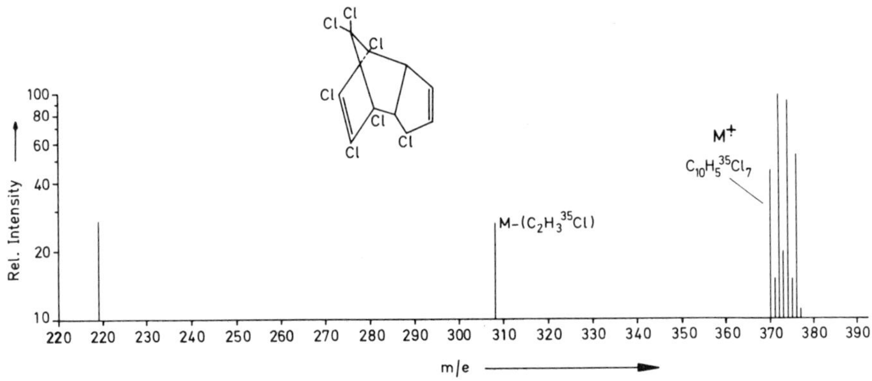

FIG. 18. FD mass spectrum of Heptachlor. Photographic recording on vacuum-evaporated AgBr plate, high resolution; emitter temperature ∿200°C [46].

rel. int., $[M + 1]^+$), m/e 361 (3% rel. int., $[M - OH]^+$), and m/e 343 (100% rel. int., $[M - Cl]^+$). Figure 17 shows the FD-MS recorded at room temperature, with electrical detection and low resolution (modified Varian MAT-CH4 mass spectrometer) [68]. It consists exclusively of the molecular ion group; fragments are not detected. Under the above-specified conditions, FD spectra can be obtained for other pesticides of the dimethanonaphthalene type such as Dieldrin, Aldrin, and Isodrin [69] in which only the molecular ions appear.

FD-MS of insecticides of the dimethanoindene type have been published so far for Chlordene, Heptachlor, and Endosulfan. In comparing the EI and FD-MS of Chlordene [70] one obtains also, under the same experimental conditions as the FD measurements of the dimethanonaphthalene type, the typical picture of no fragments in the FD spectra and of intensive fragmentation with EI. However, unlike Endrin, Chlordene does not desorb optimally at room temperature, but desorbs at a slightly raised emitter temperature (8 mA emitter heating current ≈50°C). Also, in the EI-MS, the decrease of volatility becomes apparent, because only two signals (m/e 101 and m/e 237) are recorded with more than 3% relative intensity above m/e 66. Figure 18 shown the FD-MS of Heptachlor [46], recorded with photographic detection on Ilford Q2 plates. Here one takes advantage of the integrating function of the photoplate by increasing the heating current slowly from 0 to 50 mA during the exposure time of about 20 min. At a small heating current the solvent used is first desorbed (e.g., m/e 58 acetone); with the continued increase of the emitter temperature the optimal value is reached (T*), and high molecular ion intensities are produced. Further increase of the heating current results in a thermally- and field-induced fragmentation in the molecules of the substance still left on the emitter surface. Therefore, signals, which indicate this formation of fragments, appear in the spectrum in addition to the prominent molecular ions. Since the high-temperature-activated emitters produce excellent yields of ions, the resulting ion currents are sufficient for high-resolution mass spectrometry. The accurate mass measurements for the FD-MS of Heptachlor give a signal at m/e 307.889 and this elimination can be explained as $[M - C_2H_3{}^{35}Cl)]^+$. The elemental composition $C_9H_4{}^{35}Cl_2{}^{37}Cl$ can be assigned to an additional signal at m/e 218.995. It must be emphasized that, with electrical detection and low resolution, no fragments can be detected in the mass range from m/e 200 up to the molecular ion group because of the slow scan (scanning speed 2-3 min/decade and paper plotter) [32]. Comparison with the EI spectrum also recorded by Damico et al. [45] shows that 60 signals are registered alone in the range from m/e 210 to m/e 380 and that the molecular ion intensity is about 3%. The base peak in the EI-MS of Heptachlor lies at m/e 100 $[C_5H_5Cl]^+$. In CI-MS the molecular ion intensity of Heptachlor at m/e 370 is observed to be only 4%; the base peak of the spectrum is formed by the $[M - Cl]^+$ signal at m/e 335 [67]. The EI-MS of another insecticide of this class, Endosulfan, can be described similarly as the spectra of Chlordene and Heptachlor. In FD-MS with photographic detection and high resolution the molecular ion forms the base peak. Even at emitter heating currents between 0 and 30 mA the only fragment group is due to SO elimination. More complicated, bridged carbocyclic pesticides, such as Kepone [69] and Kelevan [70], also adhere to the general rule of high molecular ion intensity and little fragmentation with FD-MS. The EI-MS of Kepone shows the $[M]^{+\cdot}$ ion with only 1% relative intensity [71]. On the other hand, in the FD-MS no fragment is detected between m/e 58 (the signal for the solvent used) and the molecular ion group if the spectrum is recorded photographically at an exposure time of 12 min with emitter heating current between 0 and 20 mA.

b. Chlorinated diphenylmethanes. As an example of the behavior of principal pesticide substances we shall mention the widely distributed, extensively used insecticides of the chlorodiphenylmethane type. Figure 19 shows the FD spectrum of the parent pesticide, p,p'-DDT with photographic detection at a resolution of about 15,000 (10% valley definition) [72]. With the intention to produce thermally and field-induced fragmentation, this FD spectrum was recorded at a raised emitter temperature (5 mA emitter heating current). If the FD-MS of p,p'-DDT is taken at room temperature the molecular ion group is practically all that is registered with electrical as well as with photographic detection. With EI-MS $[M]^{+\cdot}$ has a relative intensity below 5% as do the signals at m/e 317 and m/e 282 $[M - Cl; M - 2Cl]^+$. The loss of one or two chlorine atoms from the DDT molecule is indicated more distinctly with FD, but even more pronounced is the direct bond rupture between C(1) and C(2) which leads to m/e 235 (base peak with EI-MS) [73]. Since a relative intense signal is also observed at m/e 117 for $[CCl_3]^+$, a typical feature of many FD spectra becomes evident, namely, the frequent appearance of the splitting of

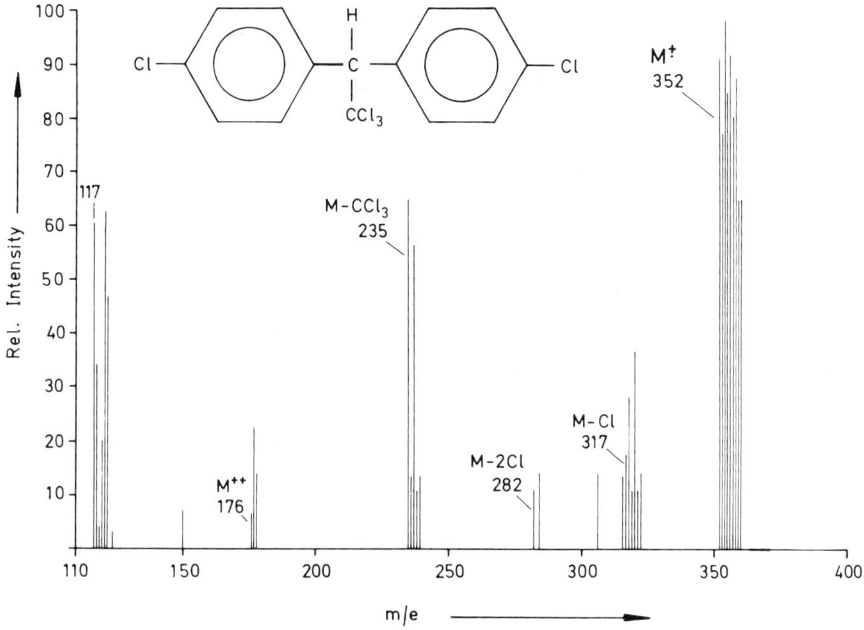

FIG. 19. FD mass spectrum of 1,1,1,-trichloro-2,2-bis (p-chlorophenyl)ethane (p,p'-DDT): emitter heating current, 5 mA; solvent acetone [72].

the investigated molecule into two complementary, charged particles. The interpretation of the spectrum is considerably simplified by these direct bond ruptures. The analytical interpretation of the data obtained is also relatively easy for nonexperts of mass spectrometry. These "one-step processes" can be more easily recognized than the complicated rearrangement and consecutive decomposition mechanisms of other ionization methods [38, 74].

Doubly charged ions, such as $[M]^{++}$ at m/e 176 and $[M - CCl_3]^{++}$ at m/e 117.5 are particularly interesting because of the theoretical aspects of their production in the high field. The investigations of Röllgen and Beckey [75-77] have shown that with FI-MS intermediary bonds of the singly charged ions are an important feature for the formation of doubly charged ions. The corresponding FD processes are complicated because of the solid layer of adsorbate. For doubly charged ions which were generated by bond rupture and/or elimination of neutral particles from the primary molecule, one finds frequently an intense signal at the one-mass-number-higher mass. Exact mass determinations indicate that a protonation in the adsorbed layer can be taken into consideration for the acceptance of the second charge (and one mass unit). This was demonstrated on sultam derivatives, for example Ref. 38. The FD measurements of Games et al. [78,79] on alkali salts of bile acids (for instance sodium glycocholate) showed doubly charged ions of a $[M + 2Na]^{++}$ composition.

If one refers to the processes of the field desorption of organic molecules, it is better to speak generally about the transfer of a cation as "cationization" [80] except for the case of "protonation." From an analytical point of view this process is especially important, because the attachment (for example, of an alkali cation to a large organic molecule with nucleophilic functions) produces stable "even-electron" ion species. Thereby the possibility of ionization and detection of the intact molecule is considerably increased [39].

In summary one can say about the pesticides described above that the molecular ion intensities obtained with EI in many cases suffice for identification of these compounds. The isotope distribution of the chlorine atoms present in these compounds constitutes an easily recognizable criterion. This holds true when the signal to noise ratio permits a high electronic amplification of the signal and in the case of investigating a pure substance. Abramson [81] was able to detect mixtures of, for instance, Heptachlor, Dieldrin, and Endrin in 100 μg/liter concentration with a combination of a mass spectrometer and a gas chromatograph (with data systems and multiple specific ion detection).

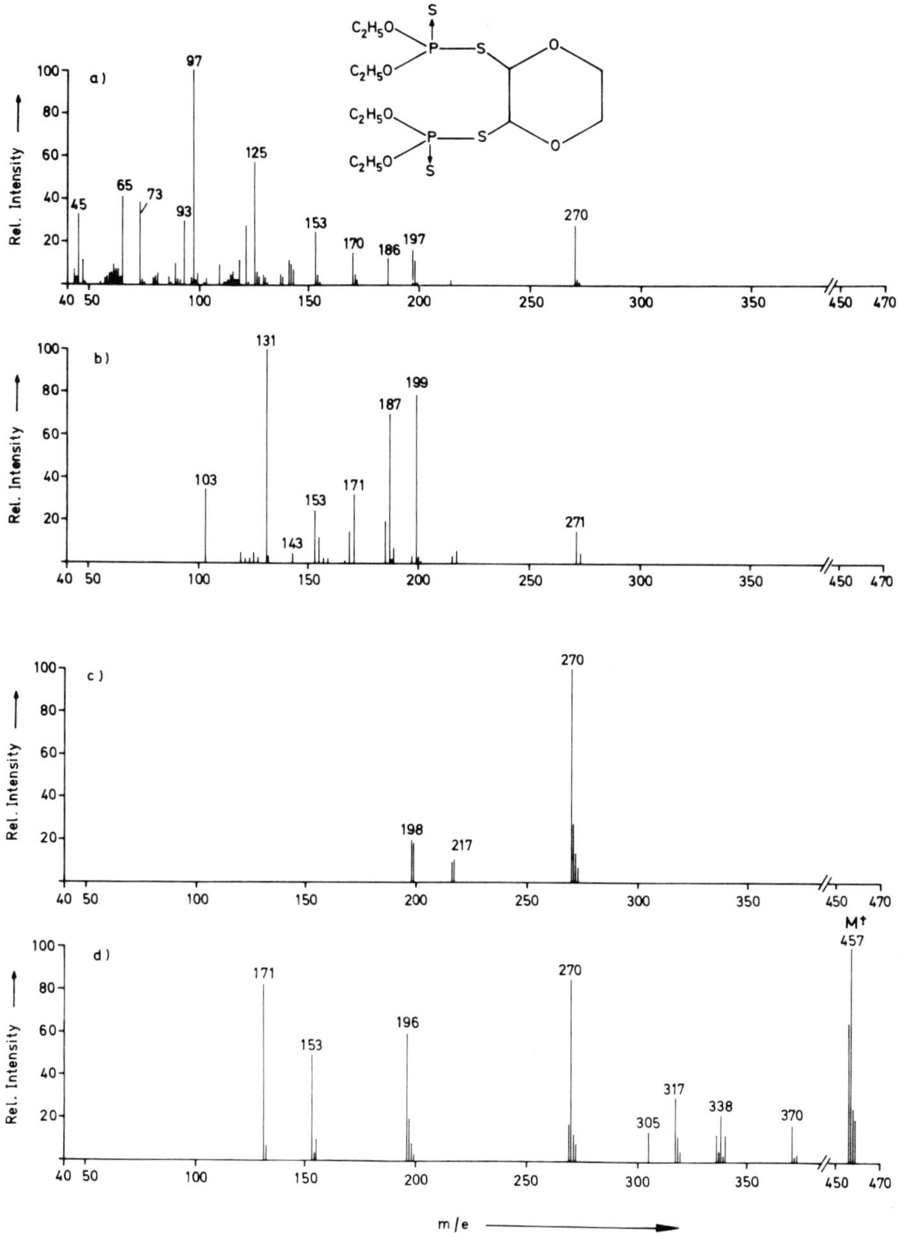

FIG. 20. (a) EI, (b) CI isobutane, 200°C, (c) FI, and (d) FD mass spectra of Delnav [55].

c. Organophosphorous pesticides. The characteristic properties of the FD method become more evident when one compares the EI, CI, FI, and FD mass spectra of several pesticides such as Delnav, Phosphamidon, and Temik. Fales et al. [55] were able to demonstrate that, indeed, the decisive feature of field desorption rests on the comparatively small thermal stress; it is not a consequence of the small energy transfer at the ion formation as in the comparison of CI/FI and EI. Since the vaporization process required in other methods is avoided with desorption in the high field, the chance to detect the ionized molecule intact is increased. In order to make this decisive point clear, we contrast in Figure 20 the EI, FI, CI, and FD spectra of Delnav. Without going further into the details of the spectra, it is seen, at first glance, that in the high-mass range a signal is registered with all ionization methods at m/e 270 (271 with CI) for the process $[M - (EtO)PS_2H]^+$. The molecular ion is detected only with FD-MS and is the base peak of the spectrum. The FD-MS of Diazinon, a contact insecticide and acaricide, in Figure 21 is an interesting spectrum for high-resolution studies [82]. The precise mass determination of the molecular ion at m/e 304.101 was performed with an error of only 0.5 millimass units ($C_{12}H_{21}O_3N_2PS$). The ions at m/e 288 and m/e 289 are probably due to the molecular ions of an accompanying compound, diazoxon. At the nominal mass 152 a doublet is recorded. One peak, accurate mass 152.051, is produced by the doubly charged molecular ion. The second signal, accurate mass 152.095, has the elemental composition of $C_8H_{12}ON_2$ and is formed by the direct bond cleavage between oxygen and phosphorous atoms. Protonation generated 2-isopropyl-4-methyl-6-hydroxypyrimidine, a very useful fragment for structural elucidation of a major building block of the parent molecule. This direct cleavage at the glycosidic [83] and phenolic [80] oxygen together with the proton transfer has been observed in a large number of FD spectra and represents a general characteristic of field desorption.

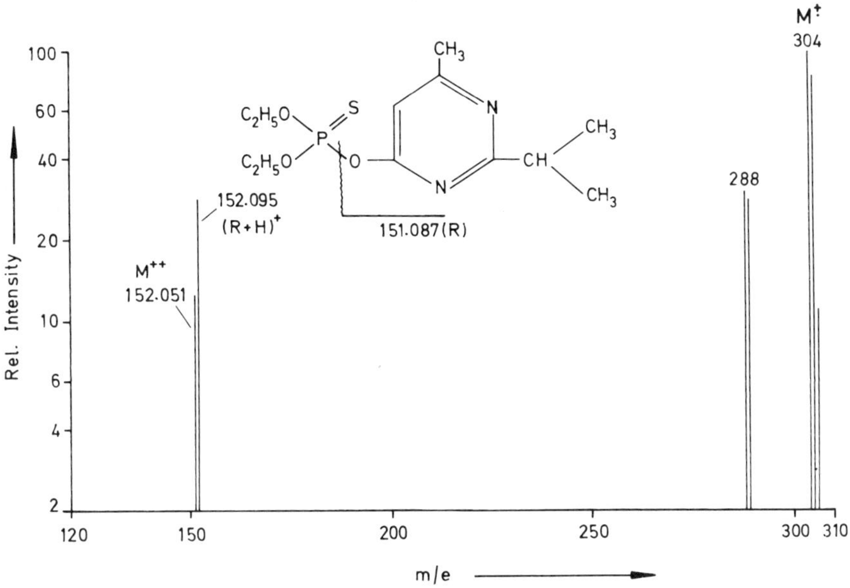

FIG. 21. FD mass spectrum of Diazinon [82].

d. Pesticidal salts. One particularly important feature of FD-MS is its capacity for the analysis of organic and inorganic salts [84, and references cited therein]. As an example we report in Figure 22 the FD spectrum of the fungicide Dodine [82]. As might be expected from the previous investigation on the formation of ions in FD-MS of salts, no molecular ion is detected. Similar to the studies of quaternary ammonium salts [85], alkali salts of bile acids [79], and other onium-substances [86,87], the cation appears with high relative abundance. The N-dodecylguanidinium ion at m/e 228.244 ($C_{13}H_{30}N_3$) is the base peak of the FD spectrum of Dodine. However, N-dodecylguanidine is displayed with comparable intensity at m/e 227.236 and yields a doubly charged ion at m/e 113.618. In view of the pronounced signal for acetic acid at m/e 60.021 and protonated acetic acid at m/e 61.029, it is assumed that a major portion of the protons are transferred to the anionic moiety of the salt, although a

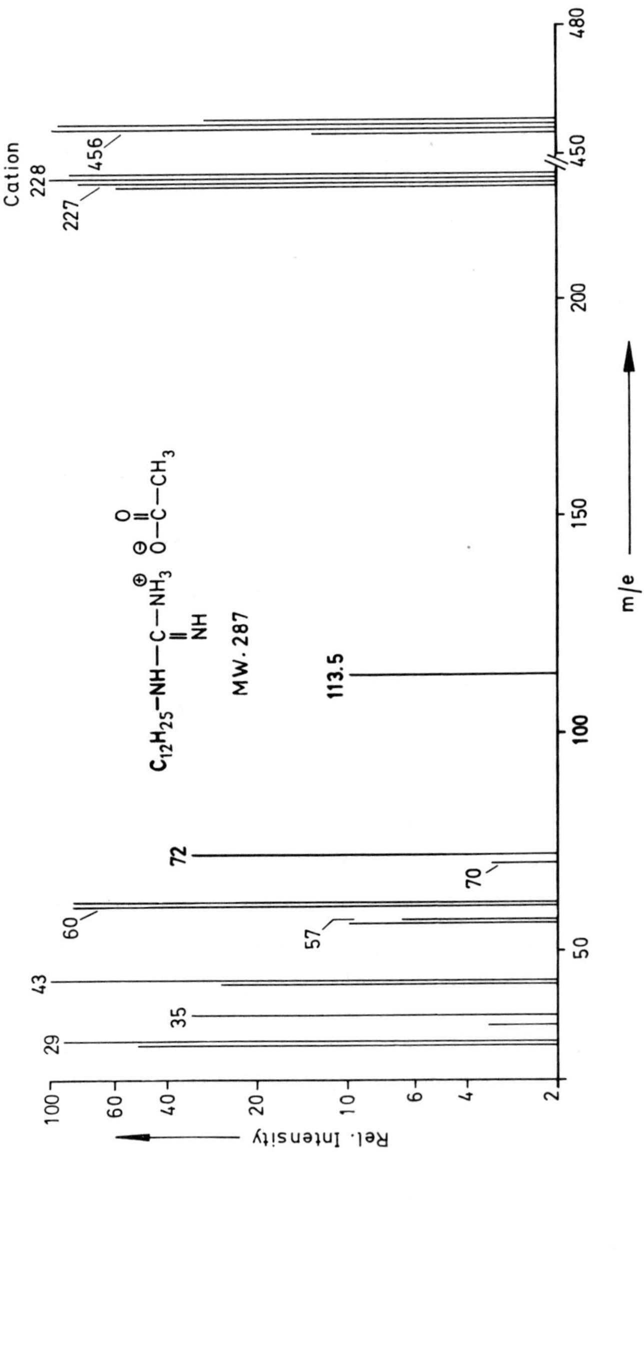

FIG. 22. FD mass spectrum of Dodine [82].

weak signal (8% log. blackening) for the guanidinium ion at m/e 60.056 (CH_6N_3) is found. Further, an intense peak at m/e 455.480 ($C_{26}N_{59}N_6$) indicates a novel type of ion in FD-MS resulting from cationization of N-dodecylguanidine by the N-dodecylguanidium ion. In addition a very weak cluster ion consisting of two cations and one anion at m/e 515 and a number of fragments from the alkyl part of the Dodine molecule are detected.

Evans et al. [88] investigated a number of salt-type pesticides with the FD technique. With the insecticide Menazon and the herbicides Paraquat and Diquat, as dihydrochloride or dihydrobromide, respectively, the cationic part of the molecules was the base peak, and little fragmentation due to the elimination of methyl, benzyl, or ethylene groups occurred.

2. *Mixtures of Commercial Pesticides*

A mass spectrometric method producing predominantly molecular ions and strongly reduced fragmentation should be well suited for the analysis of mixtures. In fact it has been demonstrated for line products of antibiotics [89], crude extracts of bile acids [79], and a combination of a parent drug with some of its metabolites [90] that in the application of FD-MS one focal point is the investigation of mixtures. Because in practice pesticides are often used as mixtures that consist of different effective environmental chemicals, one of these blends was examined with the FD technique [82].

A widely used pesticidal mixture consists of the insecticide Carbaryl and the acaricide Tetradifon. In Figure 23b the FD mass spectrum of Carbaryl is shown. The molecular ion at m/e 201.079 ($C_{12}H_{11}O_2N$) is the base peak. This is in contrast to the results obtained with EI-MS, giving 5% rel. int. and with FI-MS where 50% rel. int. for this ion is reported [45]. A prominent signal at m/e 143 is due to the direct bond cleavage at the naphtholic oxygen; in addition proton transfer leads to

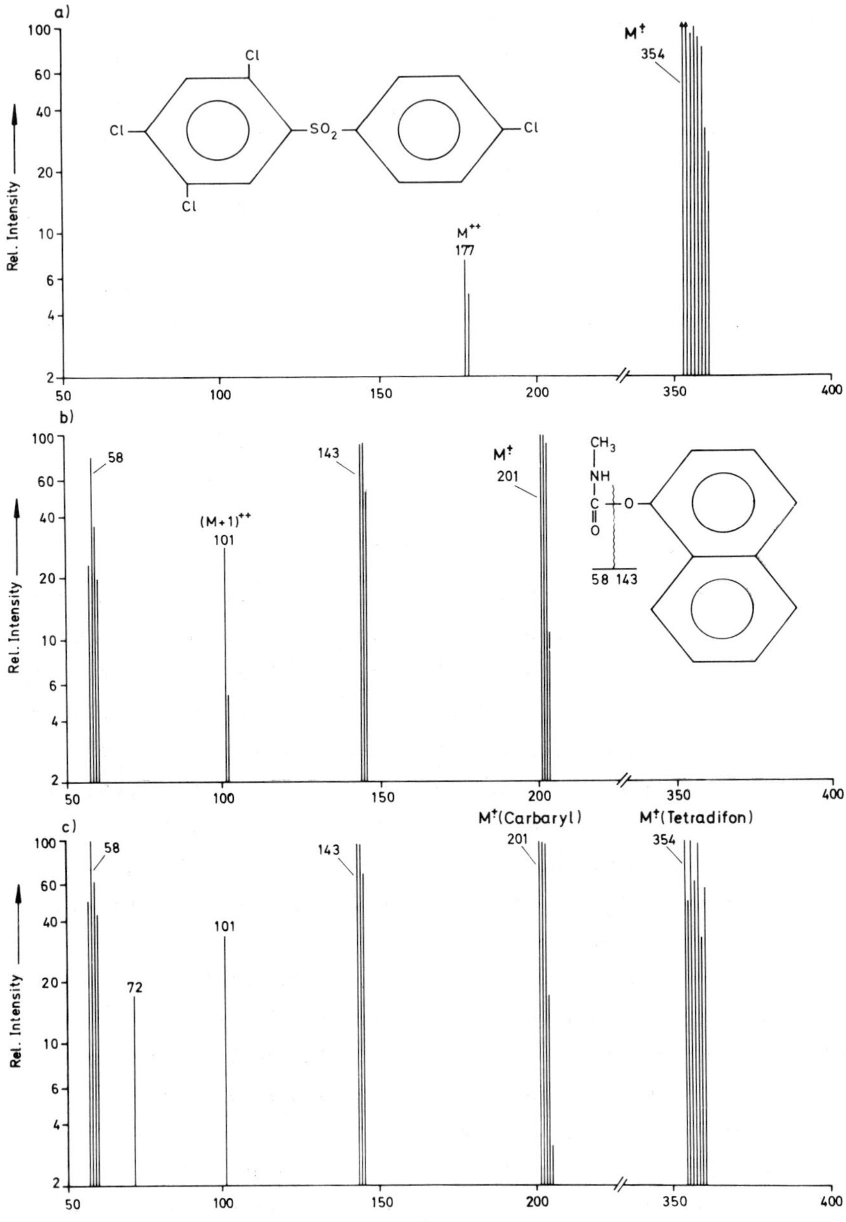

FIG. 23. (a) FD mass spectrum of Tetradifon; (b) FD mass spectrum of Carbaryl; (c) FD mass spectrum of a commercial mixture of 13 parts Tetradifon and 50 parts Carbaryl [82].

1-hydroxy-naphthalene (m/e 144). This latter peak is the base peak in EI-MS and is displayed with 80% rel. int. in FI-MS. The high-resolution FD-MS of Carbaryl reveals an interesting doublet at nominal mass 58. The upper mass at m/e 58.042 (C_3H_6O) was assigned to the solvent acetone, the lower mass at m/e 58.029 (C_2H_4ON) gives information about the other part of the Carbaryl molecule. As mentioned above this often observed direct bond fission, particularly on the linking heteroatom, accompanied by proton transfer and field ionization yield both charged, complementary parts of the examined substance molecule. Thus straightforward and easy interpretation is possible. The intense doubly charged $[M + H]^{++}$ ion at m/e 101.043 (and no $[M]^{++}$ signal) is consistent with the criteria described [40] for distinguishing between $[M]^{+\cdot}$ and $[M + H]^{+}$ ions. Further, doubly charged ions are recorded for 1-hydroxy-naphthalene at mass 72.029 $[C_{10}H_8O]^{++}$. These ions are useful to confirm the right assignment of the singly charged species in the FD spectrum .

The second component of the mixture, Tetradifon, a relatively volatile substance, showed no fragmentation in its FD spectrum (Fig. 23a). The molecular ions produce saturated blackening on the photoplate so that the $[M + 4]^{+}$ ion was chosen as base peak. Except for the molecular group only two doubly charged ions were recorded at m/e 177 $[M]^{++}$ and 178 $[M + 2]^{++}$ with very low relative intensity.

The FD spectrum of the mixture Carbaryl and Tetradifon is given in Figure 23c. Amazingly, the integrated recording on the photoplate is practically the same as adding up the spectra of the individual compounds shown in Figures 23a and 23b. The sample molecules of Carbaryl and Tetradifon desorb consecutively according to their best anode temperature in the mixture, i.e., *fractionated desorption* occurs. In this particular case no interference between the two compounds is observed.

3. Analysis of Decomposition Products of Pesticides

After having demonstrated some typical characteristic properties of FD-MS, with the example of widely distributed *primary* substances of pesticidal effect, the characteristic features of the method for the detection of *metabolic* and *nonmetabolic* decomposition products of these substances are also to be elucidated by way of example. The following question should be asked: *Why is FD mass spectrometry so especially well suited for the investigation of metabolites?*

1. Initial compounds, pesticides as well as drugs, are, according to experience, transformed into substances of higher polarity during their metabolic pathway (e.g., in higher organisms by means of increasing solubility in water). The polarity thus increases and the volatility of the intact molecule accordingly decreases, particularly when hydroxyl and carboxyl functions are introduced. With increasing metabolization, the initial substances become less suited for direct investigation with a GLC-MS combination. A coupling or combination of liquid chromatography (LC) and FD-MS, however, would evidently make an ideal analytical tool.
2. Since the decomposition products of living organisms are produced together in very small quantities, the detection limit becomes of a decisive factor. The upper and lower detection limit of FD-MS is, at present, between 10^{-6} and 10^{-11} g, depending, for example, on the type of substance and the type, resolving power, and sensitivity of the mass spectrometer. It must be emphasized, however, that these statements refer to those quantities of the substance which are adsorbed on the surface of the FD emitter. The manipulation of such small sample quantities, or rather the conveyance of the sample solution on the emitter, is subject to difficulties which also appear in other methods for micro- and submicroanalysis. True, other mass spectrometric methods also work with the stated sample quantities; however with FD, there is no necessity of derivatization. Thus, time and labor are saved, and, in addition, possible sources of errors are avoided during the introduction of protective groups.
3. The samples to be analyzed in the framework of metabolism studies are almost exclusively *mixtures*. The special ability of field ionization for the analysis of mixtures has been described in Section V B. The considerable capacities of field desorption mass spectrometry for the analyses of compounds of low volatility have been demonstrated for a wide

variety of single substances and a few mixtures. Even mixtures of very complex composition yield simplified spectra, since the single components appear predominantly as molecular ions in the FD spectrum and fragmentation only plays a minor part.

4. Pesticides and pharmaceuticals are often excreted not as free compounds but as conjugates with amino acids, peptides, sugars and sugar derivatives, organic and inorganic acids, etc. For just such substances it has been demonstrated with FD-MS by means of model compounds that it is eminently suited for this type of analytical problem.

To point 1: Decomposition products of pesticides which carry carboxylic groups, such as hexachloro-5-norbornene-2,3-dicarboxylic acid [46] (from Isodrin), Kelevanic acid [70] (from Kelevan), photoaldrin-dicarboxylic acid [72] (from p,p'-DDT), have been investigated with EI and FD-MS for comparison. In the same way metabolites were measured that contain hydroxyl groups such as 2,2-bis-(p-chlorophenyl)hydroxy-methane [72], 1-hydroxychlordene [70], Aldrin-4,5-cis/trans-diol [46], and Isodrin-4,5-cis-diol, which is shown in Figure 24. This FD spectrum shows particularly clearly the small thermal stress with field desorption. Despite the relatively high intensity of the ion current required for the obtained high resolution of about 28,500 (10% valley definition), the obvious thermal process of water elimination is not observed. Moreover, it is important to show that, with photographic detection, field desorption yields values of resolution which are of the same order of magnitude as other methods. The CEC 21-110 B mass spectrometer which was used gave a resolving power of about 32,000 for EI with a commercial ion source and of about 30,000 for FI with a special ion source [37].

To point 2: The detection efficiency of the FD method was discussed in Section IV.C.3.

To point 3: It has been mentioned before that the FD technique is well suited for the analysis of mixtures of commercial pesticides. These mixtures, however, are technical products

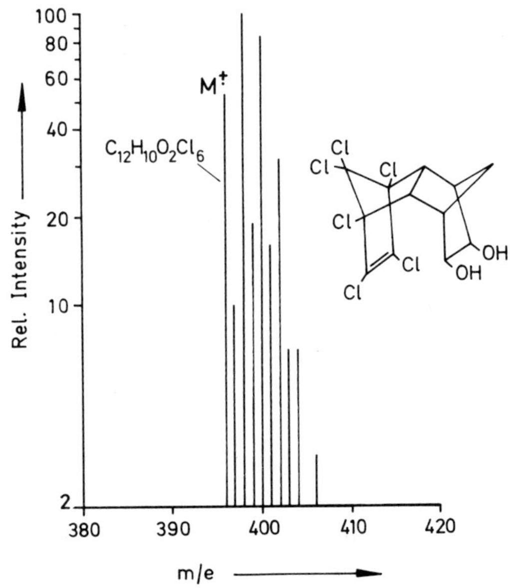

FIG. 24. FD mass spectrum of Isodrin-4,5-cis-diol, high resolution, photographic detection [46].

and their components can usually be determined by other analytical methods. For mixtures of parent pesticides together with some of their metabolites isolated from biological sources (such as body fluids and plant extractions) the amount of available material for one analysis might be a deciding parameter. Except in a preliminary FD-MS approach to identify a parent drug and some of its metabolites without derivatization in a test mixture [90], little is known about the potential of the method in mixture analysis of compounds in a wide polarity range. In order to explore the analytical resolution of the FD technique for a combination of a parent pesticide and some of its metabolites or nonmetabolic decomposition products, a mixture of p,p'-DDT and five substances that are known metabolites or decomposition products of the insecticide has been examined by FD-MS [82]. The spectrum of this synthetic mixture of pure substances is shown in Figure 25. The spectrum consists almost exclusively of the molecular ions of the individual components. Accurate mass measurements allowed the determination of the molecular weight with an average error of ±2 millimass units. In view of the small samples consumption (10^{-6} to 10^{-11} g adsorbed on the FD emitter) and the high sensitivity, the method is well suited for the analysis of crude or partially purified extracts of pesticides and pesticide metabolites. Exact quantitative measurements of mixtures with FD-MS will be mentioned at the end of this chapter. The qualitative agreement of the FD spectra with the composition of mixtures was confirmed for a synthetic blend of antibiotics [89] and for different steroidal conjugates [91].

To point 4: Here the most extensive investigations with FD-MS are known [92]. Since the mentioned compounds are not only interesting in connection with the analysis of metabolites and as conjugates, but are of general importance in biochemistry and medicine, they will be treated in detail in Section VI. B.

4. *Other Environmental Chemicals*

Considerable efforts are made to synthesize pesticides that possess short persistence and have no toxic effects on the human organism. Natural products such as insect growth regulators and sex attractants have been used successfully in the control of pest insects [93]. Since these substances are highly specific to the target animal, are active in relatively low concentrations, and may be easily degraded, it is expected that they offer an ecologically sound method of insect control.

In view of these facts the application of analytical methods for environmental protection has been directed to substances that are in general produced in large quantities, reach the environment (air, water, soil) in large amounts, and have toxic effects. The suitability of FD-MS for the detection and identification of these environmental substances has been demonstrated [95] and will be explained in the following for two classes of model compounds.

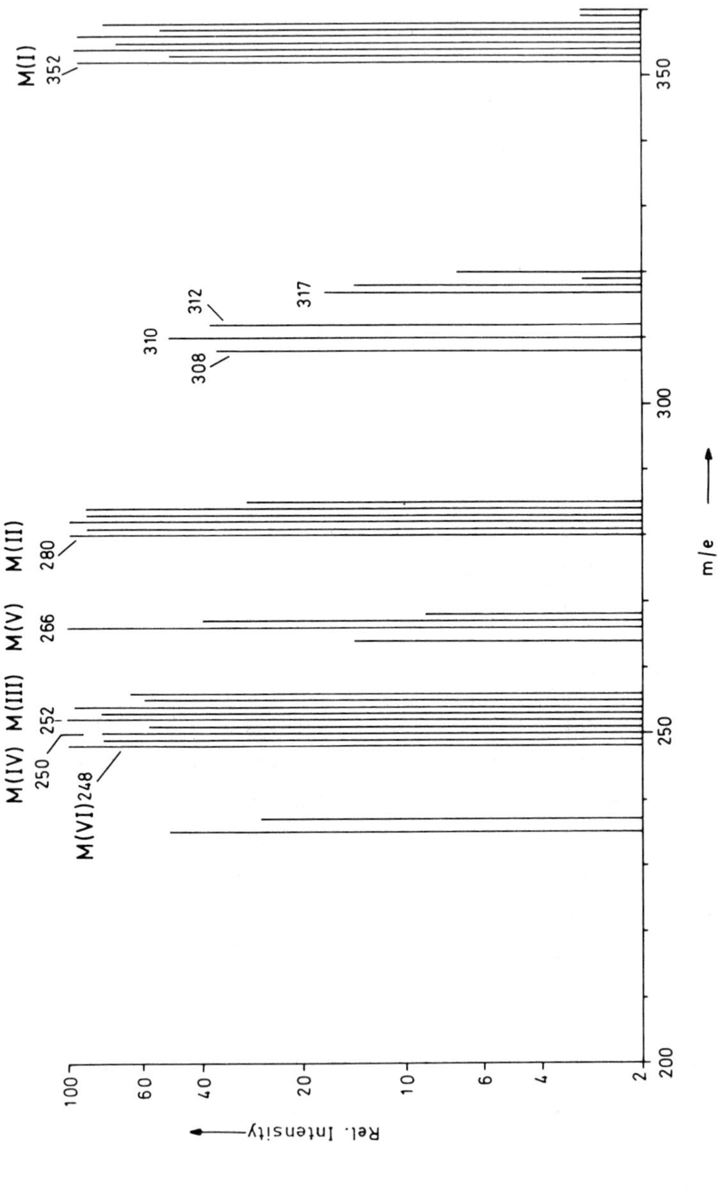
M(I)
352
310
312
308
317
M(II)
280
M(V)
266
M(III)
252
M(IV)
250
M(VI)248
200
250
300
350
m/e
100
60
40
20
10
6
4
2
Rel. Intensity

FIG. 25. FD mass spectrum of a synthetic mixture of p,p'-DDT (I) and some of its metabolites or nonmetabolic decomposition products; 2,2'-bis-(4-chlorophenyl)ethanecarboxylic acid (II), bis-(4-chlorophenyl)hydroxymethane (III), 4,4'-dichlorobenzophenone (IV), 1,1-bis-(4-chlorophenyl)ethylene (V), 2,2-bis-(4-chlorophenyl)-1-hydroxyethane (VI). Concentrations $\sim 10^{-3}$ M, the best total emission is observed between 5 and 12 mA [82]. Owing to the difficulties of establishing an ion response curve and the limited dynamic range of the photoplate, the photographically recorded FD mass spectra often do not reflect the theoretical isotope ratios (e.g., ^{35}Cl, ^{37}Cl).

a. Chlorinated paraffins. The chlorinated paraffins are technical products that fulfill exactly the three conditions above. The analysis of these complex mixtures of compounds, which are of different chain length and degree of chlorination, has been found to be very difficult by IR-absorption spectroscopy or NMR spectroscopy, particularly when only trace amounts are available [94]. EI mass spectrometry does not yield very useful spectra as in almost all cases the $[CCl_3]^+$ fragment is the most intense ion detected. The FD spectrum of a mixture of long chain chlorinated (70%) paraffins† is shown in Figure 26. The spectrum of this widely used technical product was recorded with high-resolution FD-MS. The structures of some prominent ions were tentatively assigned according to the found elemental composition. It is noteworthy that no ions above m/e 197 were detected. The occurrence of the $[CCl_3]$–group in all peaks above m/e 117 is significant for the strongly chlorinated product. The FD spectrum of a viscous paraffin containing only 42% chlorine is contrasted in Figure 27. With the same procedure for the photographic detection and the time/temperature program for the FD emitter, a completely different spectrum is obtained. Strikingly, there is a large gap in the spectrum between m/e 132 and m/e 376 where no ions are observed. A large number of prominent peaks appear in the higher-mass range between m/e 390 and m/e 556. At present it is difficult to establish a correlation between the observed ions and the structure of the two differently chlorinated paraffins, but the spectra can easily be used as fingerprints for the recognition of these two species. The differences in the spectra are so significant that the detection of trace amounts in prepurified extracts seems feasible (10^{-6} g were consumed in each FD measurement).

---

†We thank Professor O. Hutzinger, University of Amsterdam, for his suggestion to examine these compounds and his generous supply of samples of chlorinated paraffins.

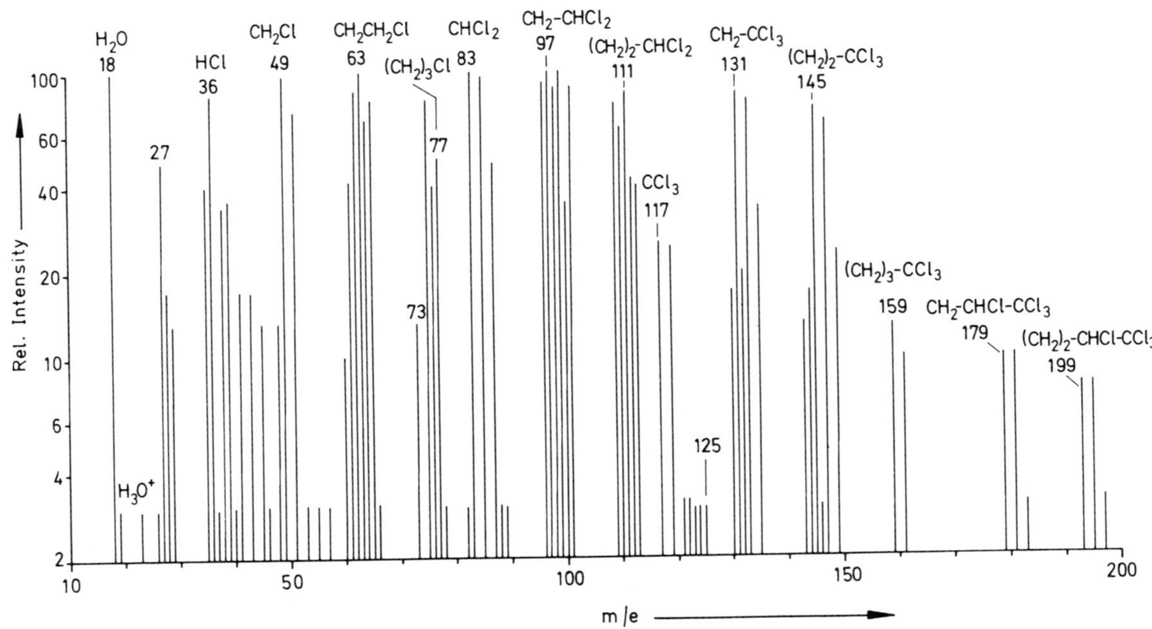

FIG. 26. FD mass spectrum of a mixture of 70% chlorinated paraffins (Cereclor 70); experimental conditions as described in Figure 27, best emission observed at 20 to 22 mA. Solid compound, solvent acetone.

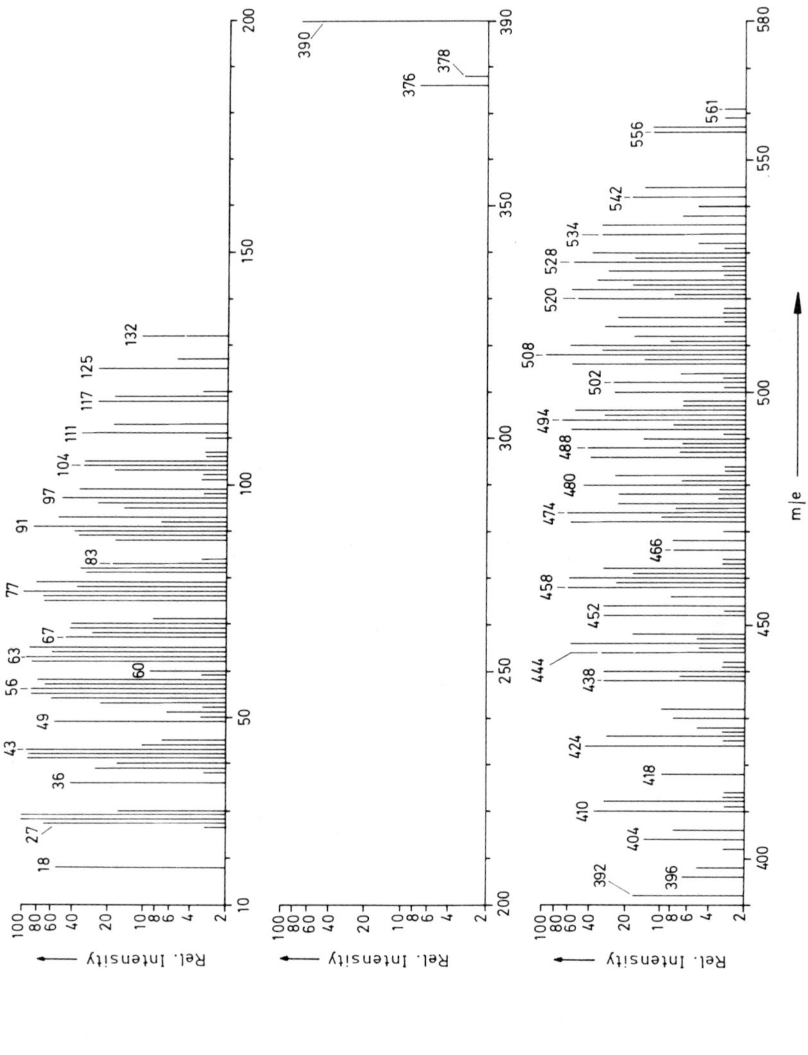

FIG. 27. FD mass spectrum of a mixture of 42% chlorinated paraffins (Cereclor 42); photographic detection; exposure time, 18 min, best emission observed at 18 mA [32]. highly viscous compound, no solvent was used.

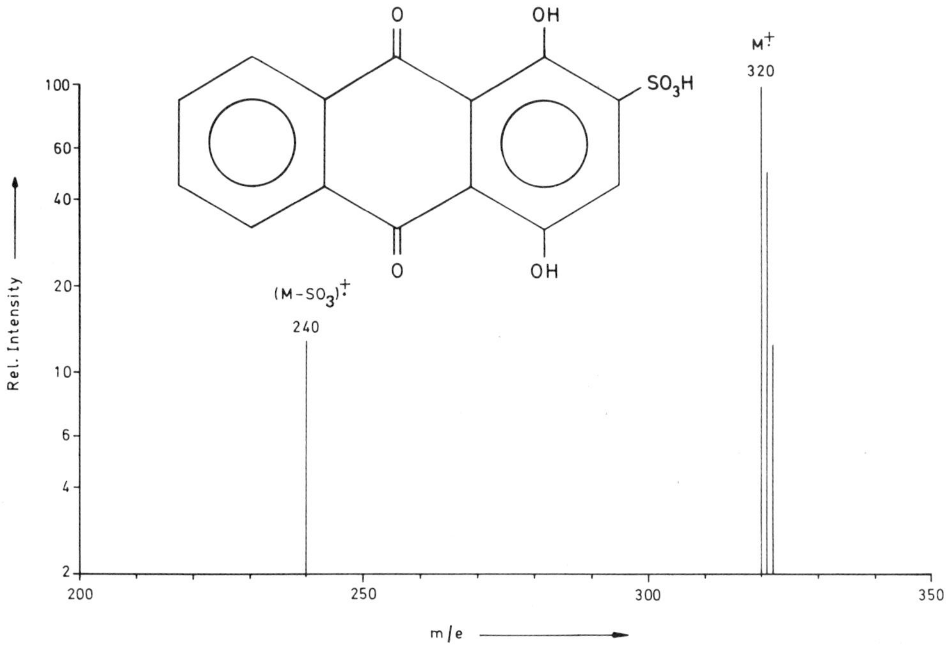

FIG. 28. FD mass spectrum of 1,4-dihydroxy-2-anthraquinone-sulfonic acid [96].

b. Dyes. Out of a wide variety of chemicals that are used as dyestuffs, we have concentrated our work on the FD analysis of free sulfonic acids and sulfonates [96]. The $-SO_3H$ or $-SO_3Na$ grouping is often purposely introduced to promote the necessary solubility properties. Technical dyestuffs are manufactured in large quantities and consequently are released into the environment. Since a large number of the metabolic and nonmetabolic products, e.g., aromatic hydroxyl derivatives, have been shown to have toxic effects, one can regard dyestuffs as environmentally dangerous chemicals. The FD spectrum of 7-iodo-8-hydroxyquinoline sulfonic acid, a disinfectant and antiseptic, has been reported previously [72]. It shows only the molecular ion and no fragmentation. In Figure 28 the FD-MS of 1,4-hydroxyanthraquinone sulfonic acid [96] is displayed. This compound is a starting material for the synthesis of a dyestuff or a possible metabolic product rather than a dyestuff. It serves, however, as an example of how FD-MS yields useful spectra of free aromatic sulfonic acids. The molecular ion is again the base peak of the FD spectrum and although two free hydroxyl functions are present in the molecule, no loss of water is observed. With photographic detection at the best anode temperature, only a loss of SO_3 from the parent molecule is recorded, which presumably is accompanied by a proton transfer to the aromatic ring system. As is often observed in the FD spectra of sulfonic acids, a signal for $[SO_2]^{+\cdot}$ is detected in the lower-mass range.

It has been demonstrated with pesticidal salts that one essential feature of FD-MS is the ability to analyze nonvolatile substances such as large organic salts. The sulfonates investigated generally showed high relative intensities for the cluster ions. The $[M + K]^+$ ion was the only ion obtained by field desorption of 1-hydroxy-2-methoxybenzene sulfonic acid potassium salt [80]. The $[M + Na]^+$ clusters were reported for sodium acetate [97] and sodium benzene sulfonate [98]. In many cases the signal for the molecular ion is completely absent or of very small relative intensity for salts. This is in sharp contrast to the usually high relative abundances of the molecular ion in FD-MS. This is particularly true when pure salts are investigated from solutions in aprotic solvents. The FD spectrum of 5,6-hydroxybenzene-1,3-disulfonic acid disodium salt is shown in Figure 29. In the mass range recorded from m/e 18 to m/e 560, the base peak is formed by $[M + Na]^+$ ions. Except for a weak peak for $[(M - H_2O) + Na]^+$ due to cationization of the product formed by the loss of one water molecule, the only intense signal is exhibited by the doubly charged ion at m/e 180 for $[M + 2Na]^{++}$. This type of ion also has been reported for

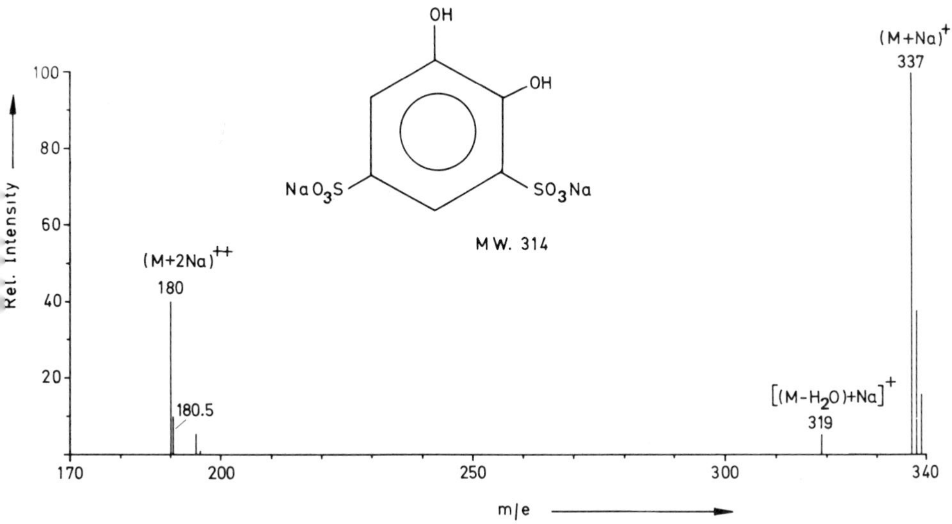

FIG. 29. FD mass spectrum of 5,6-dihydroxy-1,3-benzenedisulfonic acid disodium salt (Tiron).

the alkali salts of bile acids [79] and sodium salts of glucuronic acids [84].

To solve the problems in environmental pollution, there is a need for reliable, sensitive, and selective analytical methods for the detection of toxic environmental chemicals. The positive identification of parent pesticides, pesticidal salts, mixtures of pesticides, mixtures of pesticides and some of their metabolites, as well as involatile environmental chemicals by FD-MS has been demonstrated in the examples of model compounds. It is to be expected that the method will expand rapidly into the field of environmental research and yield useful information that cannot be obtained by other analytical methods.

B. Application in Biochemistry

1. *Sugar and Sugar Derivatives*

a. Oligosaccharides. The first substance measured with FD-MS was a monosaccharide, D-glucose [29]. This spectrum,

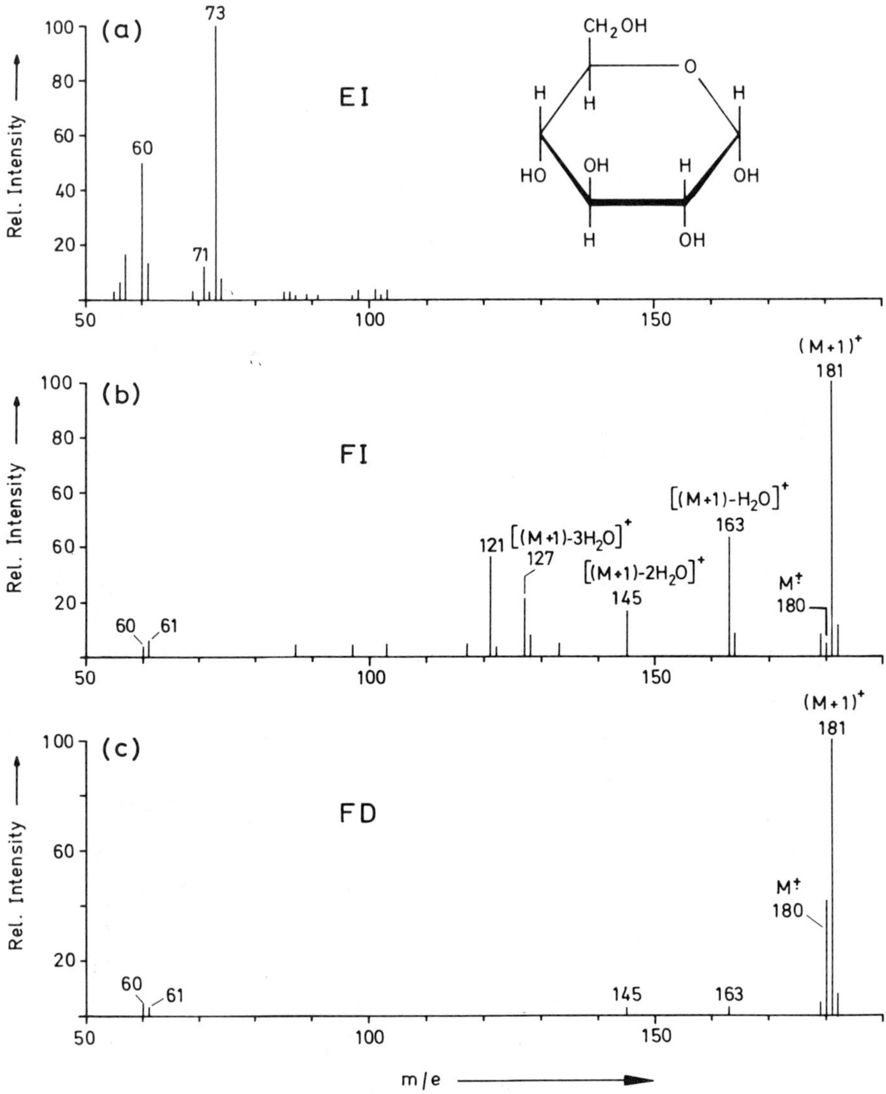

FIG. 30. (a) EI, (b) FI, and (c) FD mass spectra of glucose [29].

which is a typical example of a "one-peak spectrum," is shown in Figure 30c. It consists almost exclusively of the $[M]^{+\cdot}$ and $[M + 1]^{+}$ ions. The fragmentation does not give signals above 5% rel. int. In contrast the FI mass spectrum of glucose (Fig. 30b) displays intense ions of products that are generated by thermally induced processes such as the elimination of water from the parent molecule. The molecular ion is only of 5% rel. int., and the $[M + 1]^{+}$ is still the base peak of the spectrum as it is with FD-MS. Since the ^{13}C contribution to the [M + 1] signal is less than 0.4% of this signal, it is better described as the $[M + H]^{+}$ ion. The characteristic feature of FI-MS and FD-MS for the determination of the molecular weight of thermally labile and/or nonvolatile substances is the production of energetically favored even-electron species. In the case of glucose apparently intermolecular proton transfer plays an important role as indicated by the significant $[M - H]^{+}$ ion [1]. If the aim of the mass spectrometric analysis is the determination of the molecular weight, a comparison of the molecular ion intensities in Fig. 30c (32%) and Fig. 30b (5%) clearly shows that for this purpose FD is the more gentle and more suitable technique. Although the FD spectrum of glucose was obtained in 1969 and is a somewhat "historic" example, the decisive difference from the conventional EI spectrum is very distinct. No ion in the mass range above m/e 103 is detected in the 70-eV EI-MS of glucose in Figure 30a. The FD spectrum of a disaccharide, cellobiose, was published by Krone and Beckey [42]. The FD-MS of a tetraamino-trisaccharide is shown in Figure 31. Although this compound contains seven free hydroxyl and four free amino functions, it is the protonated molecule which gives the base signal of the spectrum. The signal with the second highest relative intensity at m/e 324 can then in turn be assigned to a direct

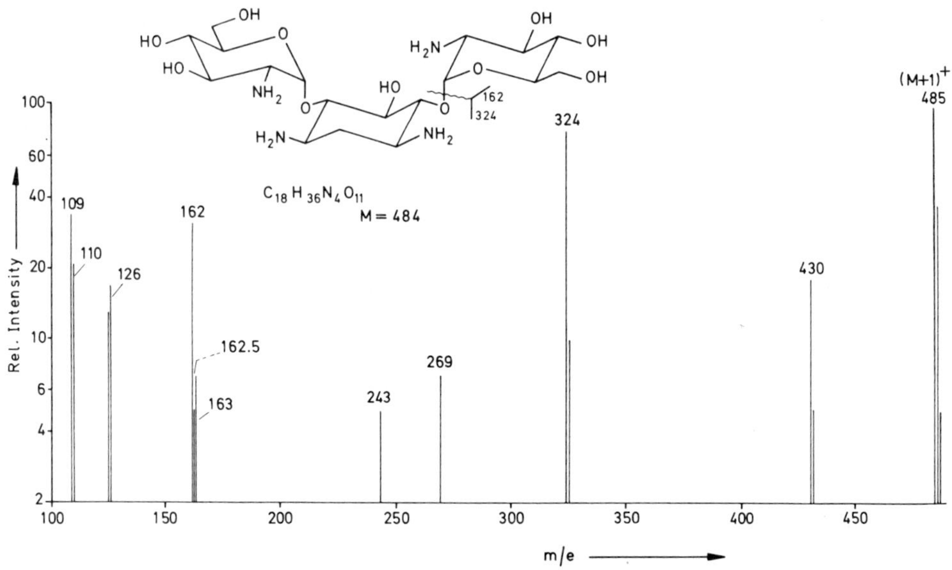

FIG. 31. FD mass spectrum of a tetraamino-trisaccharide [32].

bond rupture. Structurally related oligosaccharides with antibiotic effects such as Gentamycin and Tobramycin also yield high molecular ion intensities and bond ruptures at the glycosidic oxygen. The origin of some signals cannot be explained as yet, perhaps because these substances were isolated from natural products and thus contain admixtures. Rinehart et al. [89] have recorded the FD spectra of a number of antibiotics--Neomycin, Streptolydigine, and Novobiocin--which did not yield any molecular ions with EI and CI-MS and which showed the molecular ion as the base peak at T* with FD-MS. Moreover, qualitative indications about the composition of complex antibiotics can be obtained with field desorption as was demonstrated for instance on Streptovaricin, Filipin, and Dermostatin. To give an example the 70-eV and the 18-eV EI-MS of underivatized Neomycin B are shown in Figure 32. The spectra end in the higher-mass range at m/e 446. Surprisingly, no ion of this mass appears with FD-MS, but only

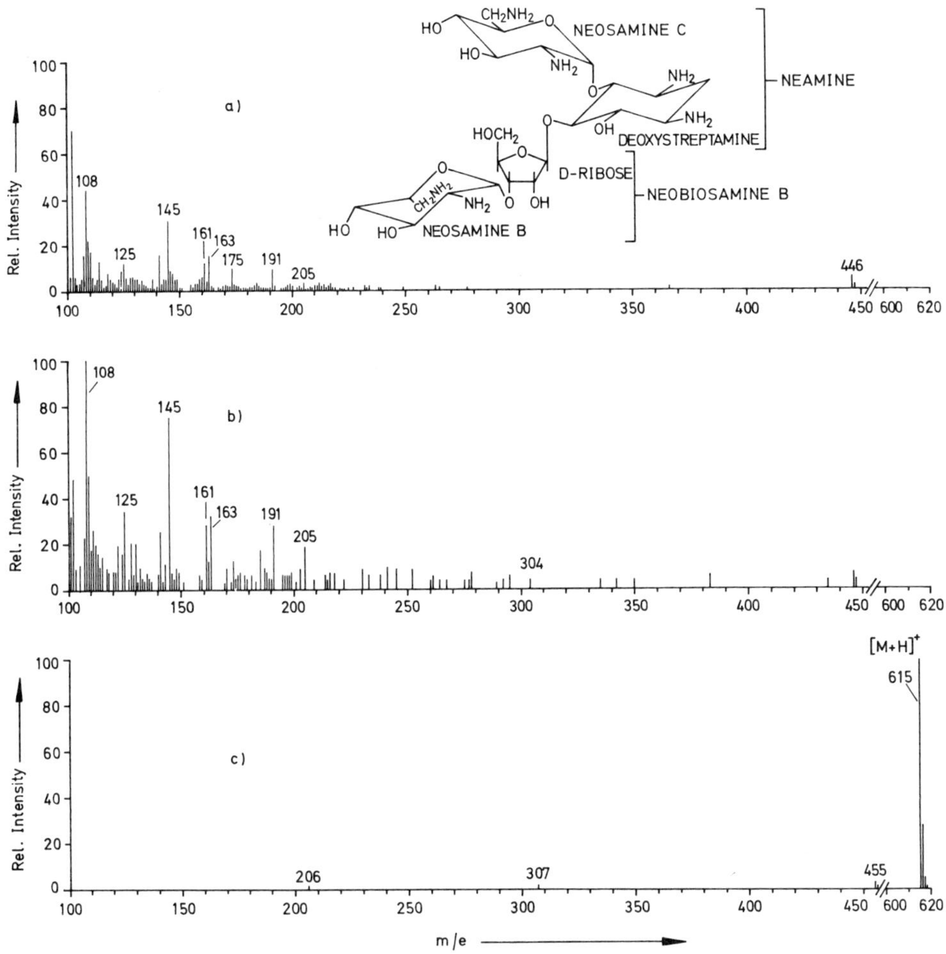

FIG. 32. (a) EI mass spectrum of Neomycin B, electron energy 70 eV; (b) EI mass spectrum of Neomycin B, electron energy 18 eV; (c) FD mass spectrum of Neomycin B [89].

relatively weak fragments appear at m/e 206, 307, and 455, while the $[M + H]^+$ ion (m/e 615) accounts for 75% of the total ion current. To obtain a mass spectrum of this compound with EI and direct inlet, it was necessary to prepare the hexa-N-acetyl- hepta-O-trimethylsilyl derivative [100]. Not only is it difficult to prepare such a derivative, but the derivatized substance becomes more than twice as heavy as the original compound because of the protective groups. Thus the analysis by MS by conventional mass spectrometers becomes borderline.

b. Glycosides. For all spectroscopic methods, there is approximately an inverse proportionality of signal intensity and resolution. Owing to the relatively small ion currents with FD, high demands are placed on the transmission of the mass spectrometer and on the quality of the emitter used in high-resolution measurements (see Sec. IV. B. 3). The advantages of high-resolution include the exact determination of the molecular weight and therewith the elemental formulas, as well as important indications in the interpretation of the fragments. High-resolution data on some arylglycosides and on glycosides of the Coumarine and Flavone type have been published [83].
Figure 33 shows the high-resolution FD-MS of Hesperidin. Here also the aspect of the stabilization of the molecular ion through cationization is made very clear. The $[M]^{+\cdot}$ at m/e 610 is indeed recognizable, but --unusual for FD--below 20% rel. int. Cationization of the Hesperidin molecule by sodium cations generates the base peak of the spectrum. The sodium ions are formed from the unavoidable admixtures of crude material extracts (for instance, after column chromatography), and at higher emitter heating currents from the tungsten of the emitter as well. The usefulness of the $[M + Na]^+$, $[M + K]^+$ signals for the differentiation of $[M]^{+\cdot}$ and $[M + H]^+$ ions of unknown compounds must be emphasized once more at this point [40]. Since high resolution gives us the elemental composition of the fragments, one can interpret m/e 302.079 as $[A + H]^{+\cdot}$ and m/e 147.066 as direct bond rupture at the glycosidic oxygen of the terminal sugar unit. The aglycon appears also as a doubly charged ion at m/e 151.040. The EI-spectrum of the free Hesperidin gives a signal for the aglycon (m/e 302) with 76% rel. int., the base peak lies at m/e 157. The molecular ion at m/e 610 is not detectable.

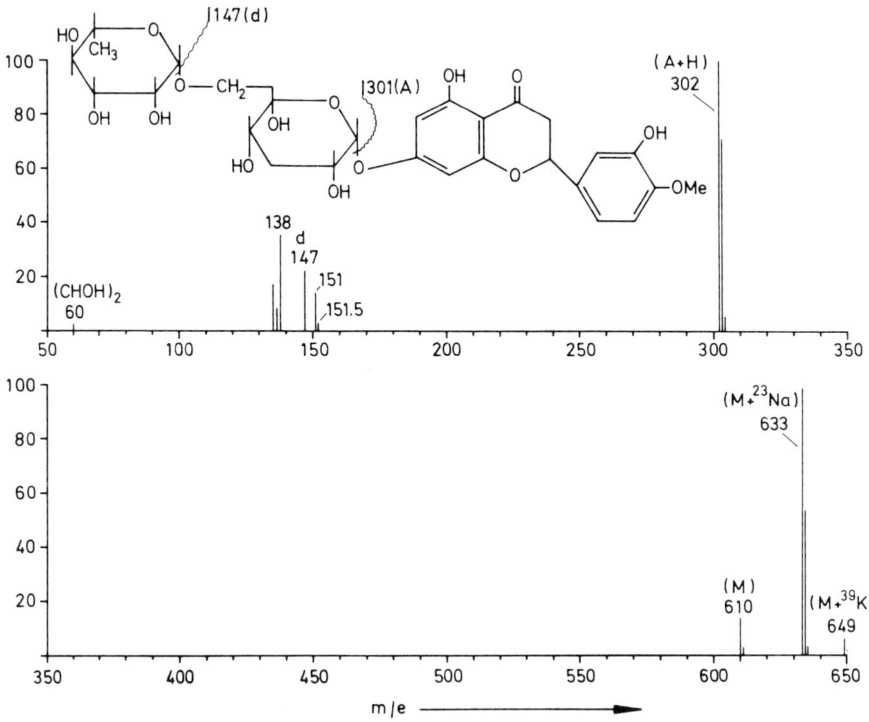

FIG. 33. FD mass spectrum of Hesperidin [83].

Investigations on stereoisomeric glycosides with aromatic aglycon, which are used as test substances for the detection of enzymes (substrate for glucosidases), showed that the FD method gives distinct differences between α- and β-linked glycosides. The comparison of EI with FI spectra showed that the most simple spectra containing the greatest differences for stereoisomers can be obtained with FD-MS [101]. A stereochemical study by EI and FD of the photodimers of cinnamic acid and related compounds showed that FD-MS allows a clear differentiation between head-to head and head-to-tail linked isomers [102]. Thus, the application of FD-MS to problems in stereochemistry appears to be promising.

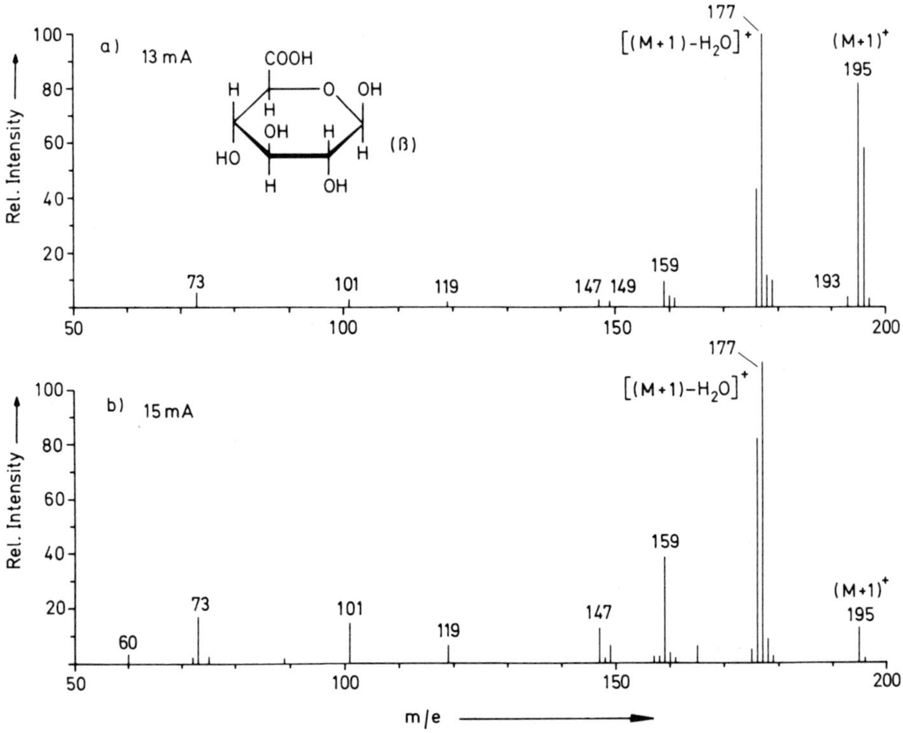

FIG. 34. FD mass spectra of glucuronic acid: (a) 13 mA emitter heating current, (b) 15 mA emitter heating current, electrical recording, solvent water [99].

c. Glucuronides. In view of the importance of the basic compound glucuronic acid for the metabolic pathway of endogenous substances and the fact that the metabolites of drugs and pesticides are often excreted as conjugates, the behavior of the free acid under the conditions of field desorption was examined. As was the case with sulfonic acid derivatives, the increased water solubility goes hand in hand with higher polarity and thermal lability upon evaporation. Conventional mass spectrometric techniques are of limited value in the study of this compound, owing to the nature of the resulting spectra. In the

EI spectra of free glucuronic acid the molecular ions are nonexistent and high mass fragments are weak. In sharp contrast, the FD spectrum electrically recorded at 13 mA emitter heating current and at an emitter temperature slightly above T* exhibits an $[M + 1]^+$ ion of 82% rel. int. (Fig. 34a). The base peak is formed by the elimination of water from the protonated molecule. The prominence of this ion species clearly indicates that the FD spectrum was run at an emitter temperature greater than T* (∿11 mA). Since the molecular ion is not observed, this confirms the previously discussed prominence of the energetically and kinetically favored even-electron species in FD-MS. When the FD spectrum is recorded at an emitter heating current which is 2 mA higher (∿15 mA), dramatic changes in the relative intensities of the obtained ions occur. The relative abundance of the $[M + 1]^+$ ion drops from 82% to 13% and the signal due to the loss of two water molecules from the protonated glucuronic acid at m/e 159 increases from 9% to 29%. Although the $[(M + 1) - H_2O]^+$ ion is still the base peak of the spectrum, the generally enhanced abundance of the decomposition products (e.g., m/e 60, m/e 73 as in the FD-MS of glucose) indicates that a thermally and field-induced fragmentation occurs.

At this point it may be worthwhile to demonstrate once more the decisive influence of the emitter temperature on the nature of the FD spectra obtained. Note specifically the striking alteration of the spectra caused by small changes in the emitter heating current. In order to obtain reproducible spectra with the electric detection system at a constant field strength with standard emitters (10 μm tungsten wires carrying 30 μm high-temperature-activated microneedles), using the same solvent and same sample amount, the time/temperature program for desorption has to be controlled very carefully. This can be achieved either directly by the data system, or by an electronic device that generates defined and reproducible time/temperature

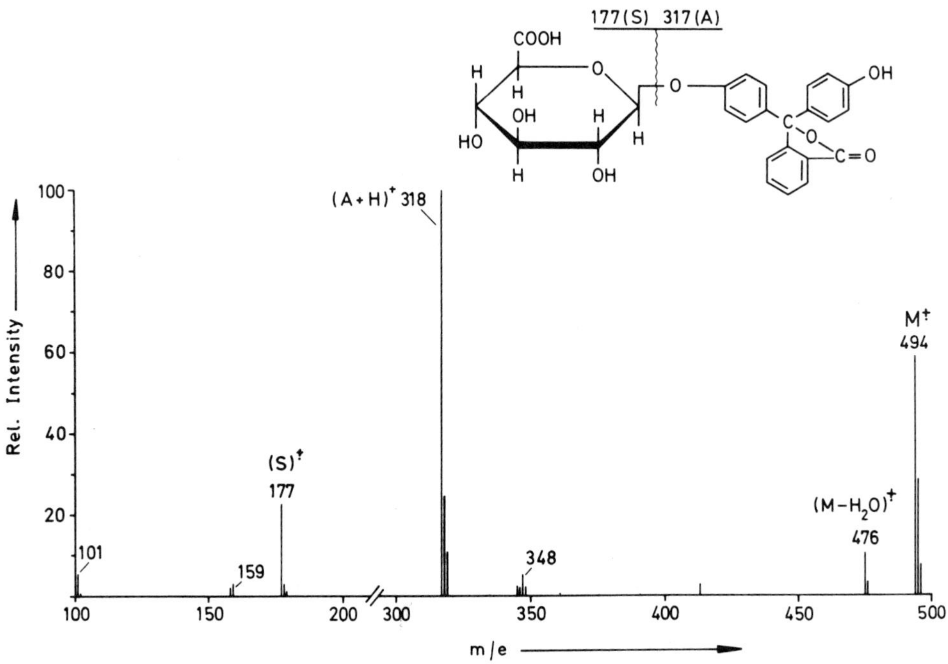

FIG. 35. FD mass spectrum of phenolphthalein-β-D-glucuronide; 12 mA emitter heating current; solvent, water; electrical detection [99].

profiles for desorption [74, 103]. Many of the biologically important derivatives of glucuronic acid are bound to the molecule at the glycosidic oxygen [C(1)]. In order to test the usefulness of FD-MS for the analysis of these substances, a spectrum of phenolphthalein-β-D-glucuronide was electrically recorded at 12 mA emitter heating current (Fig. 35). As might be expected from the FD-MS of glycosides the protonated aglycon at m/e 318 forms the base peak of the spectrum. Further, the sugar moiety (S) at m/e 177 is clearly detected. As the FD spectrum is obtained close to T*, the loss of water from the parent molecule is only 16% rel. int. These results support the supposition that the formation of ions that constitute complementary building

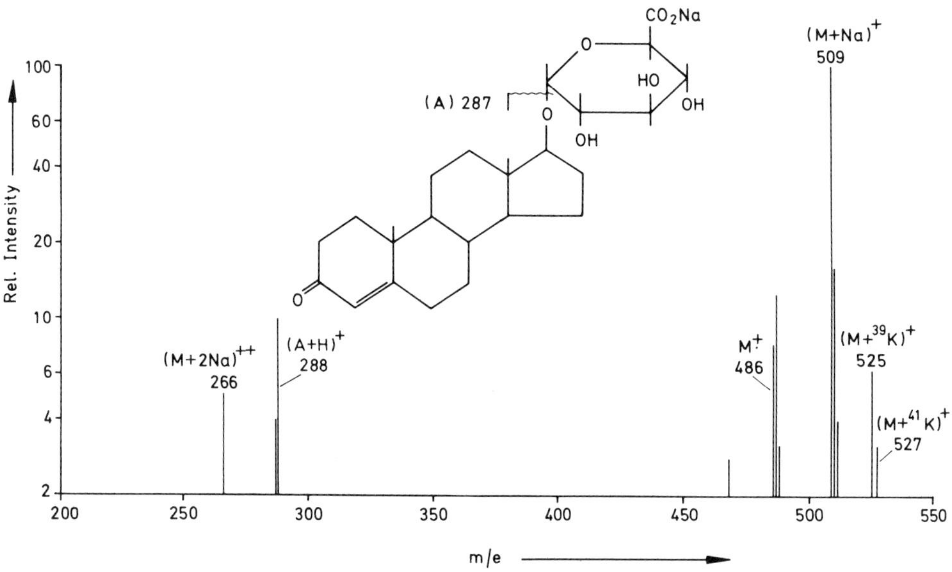

FIG. 36. FD mass spectrum of testosterone glucuronide sodium salt [84].

blocks of the parent molecule is a more general feature of FD-MS. Summarizing the FD results obtained with oligosaccharides, glycosides, and glucuronides one can say that FD-MS generally enables the molecular weight determination due to the high intensities of the molecular ions or the protonated molecules. This finding is confirmed by the investigations of free oligosaccharides and their permethylated and peracetylated derivatives by Moor and Waight [104]. Similar results are reported by Bursey and Sammons in a survey of the applications of FD-MS in carbohydrate chemistry [105]. The utility of field desorption for the analysis of an important substance class for urine extraction of hormones, the steroidglucuronides, is illustrated by the example of testosteronegluruconide. Figure 36 shows the high-resolution FD-MS of this compound. The work of Adlercreutz et al. [91] goes beyond the investigations on model substances and brings field desorption into the service of clinical chemistry and medicine. Thus oestriol-16α-glucuronide, isolated from the urine of pregnant women, was successfully detected with high molecular ion intensity. This was done without derivatization at low resolution and was supported by the FD measurement of standard synthetic compounds. It has to be said, however, that for physiological studies of endogenous compounds and pharmacokinetic investigations the quantitation of the FD results is a basic requirement. Preliminary investigations into the use of deuterated analogs of a drug for quantitative determinations of the drug by FD-MS have been reported [117].

d. Nucleosides, nucleotides. These compounds are taken into consideration, on the one hand, as forming conjugates and are thus interesting for FD analysis in an underivatized state. On the other hand, a great number of modified, biologically active nucleosides are synthesized [106], and FD-MS could give valuable information on the elucidation of structure of the compounds resulting from these syntheses. As model compounds, the important nucleosides which appear as components of RNA and DNA were measured with FI [47] and FD [37]. Thereby, it became obvious that, with the exception of guanosine, the molecular ion intensities were sufficient to identify the underivatized compounds with FI. The temperature of 250°C given in the reference for the direct inlet system evidently led to the thermal decomposition of the guanosine molecule *prior* to the transfer into the gas phase (melting point 240°C with decomposition). The FD-MS of guanosine [37] was recorded above T* at 18 to 20 mA with the intention of obtaining the structurally important, complementary fragments for the sugar part (S) and the base part (B) of the molecule. In this case the molecular ion appears with about 20% rel. int. In the same publication the strong influence of the solvent on FD was discussed. Figure 37 shows the high-resolution FD-MS of adenosine. The signals for $[B + H]^{+\cdot}$ at m/e 135.054

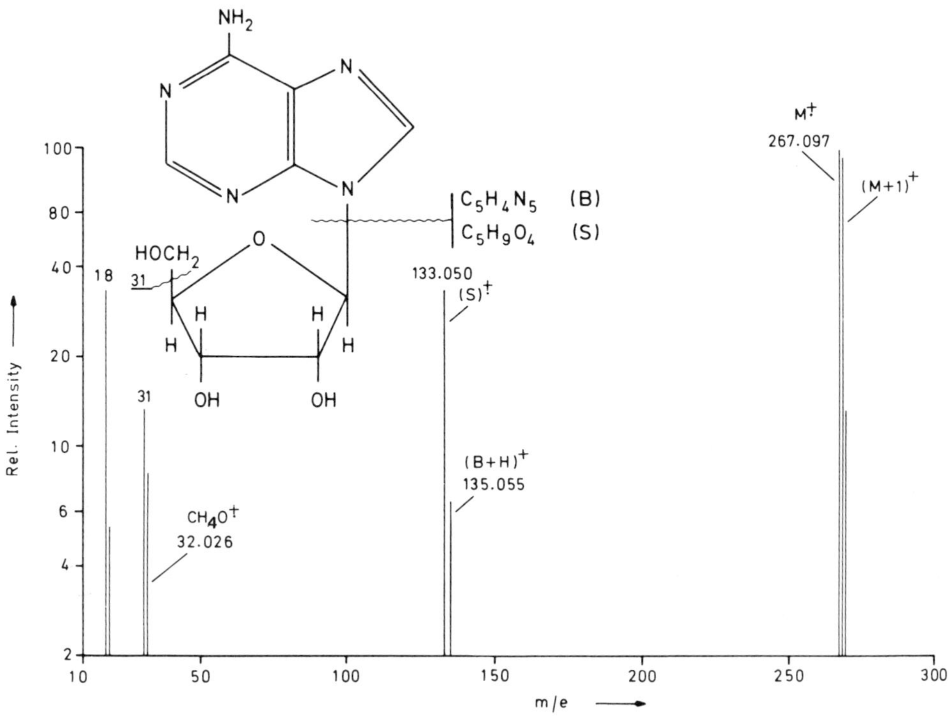

FIG. 37. FD mass spectrum of adenosine [37].

and $[S]^+$ at m/e 133.050 are characteristic as well as the base peaks for $[M]^{+\cdot}$ and $[M + H]^+$ (emitter heating current 20 mA). The 5'-adenosinemonophosphate nucleotide was the first free nucleotide to be detected with FD-MS as an intact molecular ion $[M + H]^+$ (73% rel. int., Fig. 38). The bond rupture between the phosphoric acid part and the nucleoside (N) adenosine is registered with 29% rel. int. at m/e 267.097. The remaining fragments give--as with the nucleoside--intense peaks for $[S]^{+\cdot}$ and $[B + H]^{+\cdot}$. The relatively weak signals at m/e 60.021 $[(CHOH)_2]^+$ and 61.029 $[(CHOH)_2 + H]^+$ indicate that a ribose unit is present in the molecule. At emitter heating currents above T*, e.g., up to 50 mA, protonated orthophosphoric acid is detected at m/e

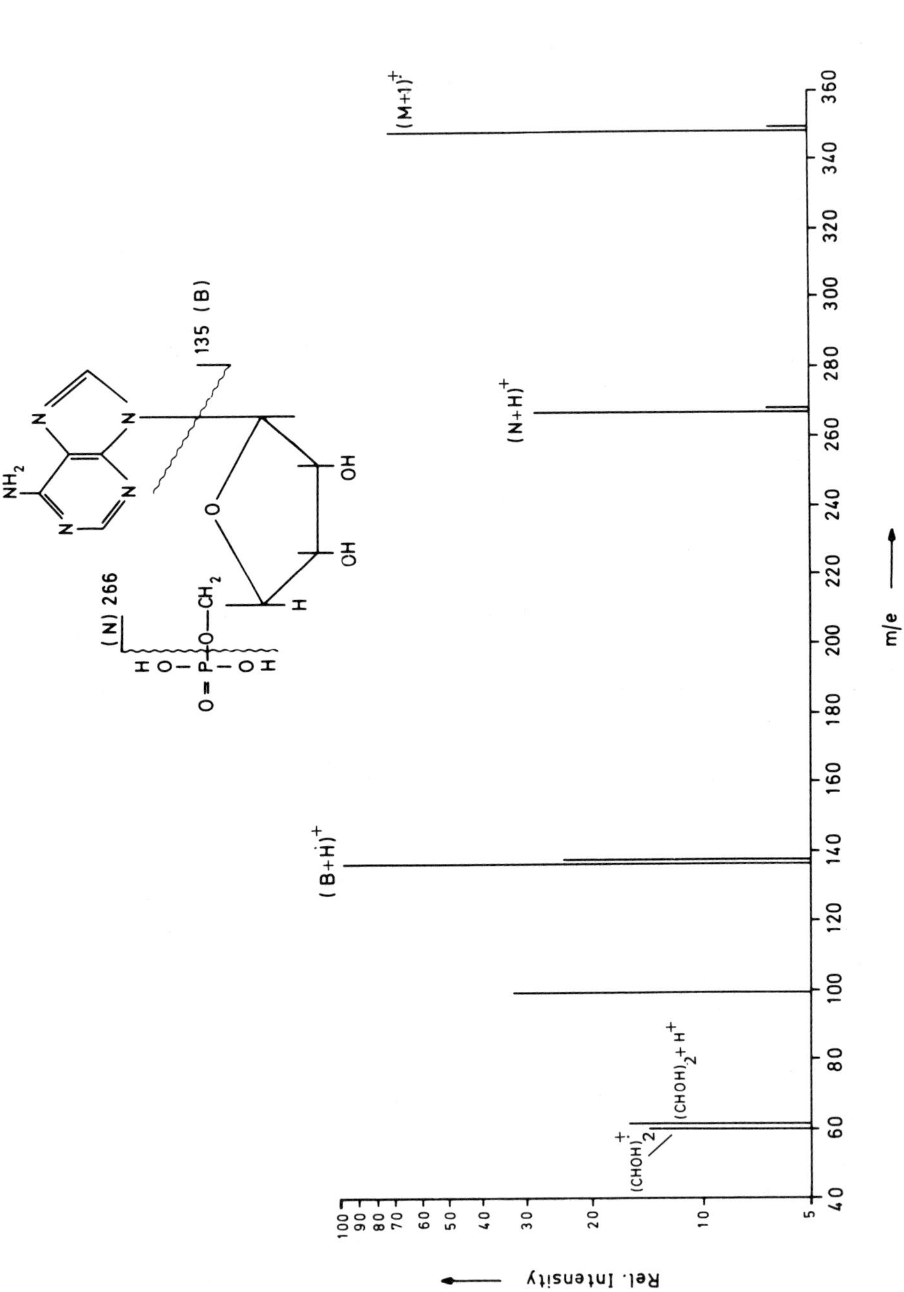

FIG. 38. FD mass spectrum of adenosine-5'-monophosphate [37].

98.985. Other free nucleotides such as 5'-thymidinemonophosphate, 5'-deoxycytidinemonophosphate, and 5'-deoxyguanosinemonophosphate gave similar spectra.

e. Vitamins. The 70eV or 12-eV EI-MS and the high-resolution FD-MS of vitamin C, ascorbic acid, have been presented in connection with the problems of detection of mutagenic compounds [72]. In this investigation the intensity of the $[M]^{+\cdot}$ ion was below 10% rel. int. with EI. With FD-MS the base peak is given by the molecular ion group $[M]^{+\cdot}/[M + H]^+$ when the spectrum is recorded at T*. Other investigators reported on the characterization of the B vitamins: pyridoxal, pyridoxamine, and their phosphates, as well as on thiamine [107]. In connection with FD spectra of the free phosphoric acid ester, it is interesting that pyridoxamine-5'-phosphate and pyridoxal-5'-phosphate as hydrochlorides give the $[M + H]^+$ ion as the base peak at T*. Further the bond rupture between the ester oxygen and the phosphorous atoms characterizes the fragmentation in a similar way to the case of FD-MS of the nucleosides, nucleotides, and glycosides.

f. Sodium salts of deoxyfluoro-D-glucose-6-phosphates. The alkali salts of sugar phosphates are at the end of this list of sugar compounds with increasing polarity. The high-resolution FD-MS of these substances [108] give the $[M + H]^+$ ion as the base peak, but also show products which result from exchange of $[Na]^+$ with $[H]^+$, elimination of water, and pyrolytic processes. These salts could only be recorded with photographic detection, since they desorb within a very narrow time and heating current interval. Tarry residues on the FD emitter at the end of the measurement (0 to 50 mA heating current, 10 min exposure time) greatly diminish the emission properties of the emitters. There are two reasons for this: First, the actual emission centers--the tips of the microneedles--are covered and second, the interspaces between the needles are no longer completely free for the adsorption of the substance to be analyzed. The observed emission of these field anodes was always sufficient for satisfactory exposure of the reference compounds [perfluorokerosene, perfluorotributylamine and tris-(perfluoroheptyl)-S-triazine] in the field ionization mode (i.e., sample supplied from the gas phase), but showed disappointing results for repeated FD measurements in actual samples.

2. Amino Acids

Several of the biologically important α-amino acids were investigated with FD-MS by Winkler and Beckey [109]. The FD spectra obtained generally give the molecular ion or quasimolecular ion with high relative intensity. In the case of arginine and cystine these ions are also clearly shown with FD although they are not detectable for either substances with EI, CI, and FI. Figure 39 shows the FD-MS of arginine. The $[M + H]^+$ ion appears with 30% rel. int., and the base peak of the spectrum at m/e 158 is marked by the loss of ammonia from the protonated molecule $[(M + H) - NH_3]^+$. In addition, a direct bond rupture occurs between the guanidyl moiety and C(5) and gives m/e 116 or m/e 117, respectively. With the α-amino acids (e.g., Glu, Val, Pro, Ser, and Cys), one frequently gets a signal at $[M - 45]^+$ which can be interpreted as $[(M + H) - COOH_2]^+$. If the FD-MS are recorded only a little above T*, one observes water elimination (e.g., with Glu, Asp, and Cys-S-S-Cys). Comparing the EI, CI, FI, and FD spectra of leucine, isoleucine, and norleucine [55], it looks as if the highest molecular ion intensity and the smallest amount of fragmentation is obtained with CI. In this case, however, compared to FD the spectra reflect the different solubilities of the amino acids during preparation of the analysis solution for FD more than the different transferred thermal energies.

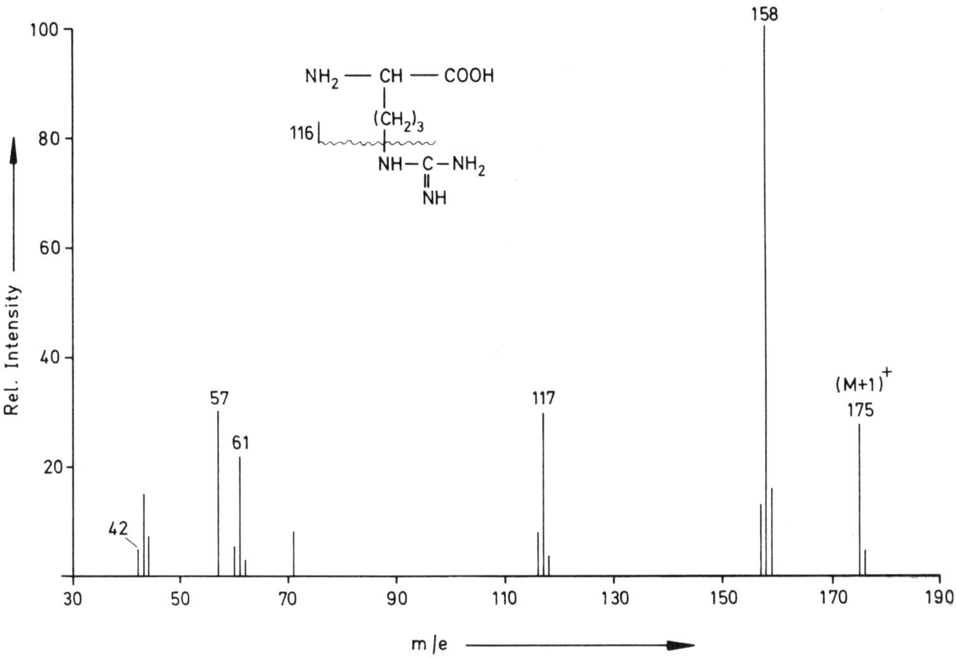

FIG. 39. FD mass spectrum of arginine [109].

3. *Peptides*

A dipeptide (glycylglycine [110]) was the first example that underivatized peptides give complete mass spectra with FD-MS in which the molecular ions form the base signal using 18 to 24 mA emitter heating current. In a subsequent FD investigation, Winkler and Beckey [111] successfully recorded the spectra of partially protected and free oligopeptides with three to nine units of amino acid (e.g., arginine and histidine), not only with high molecular ion intensities, but also with partial information about the sequence. Figure 40 shows the FD-MS of Ac-Gly-Arg-Arg-Gly-OMe as an example of these results. Although the N-terminal end of the tetrapeptide is acetylated and the C-terminal carboxyl group methylated, the base intensity of the

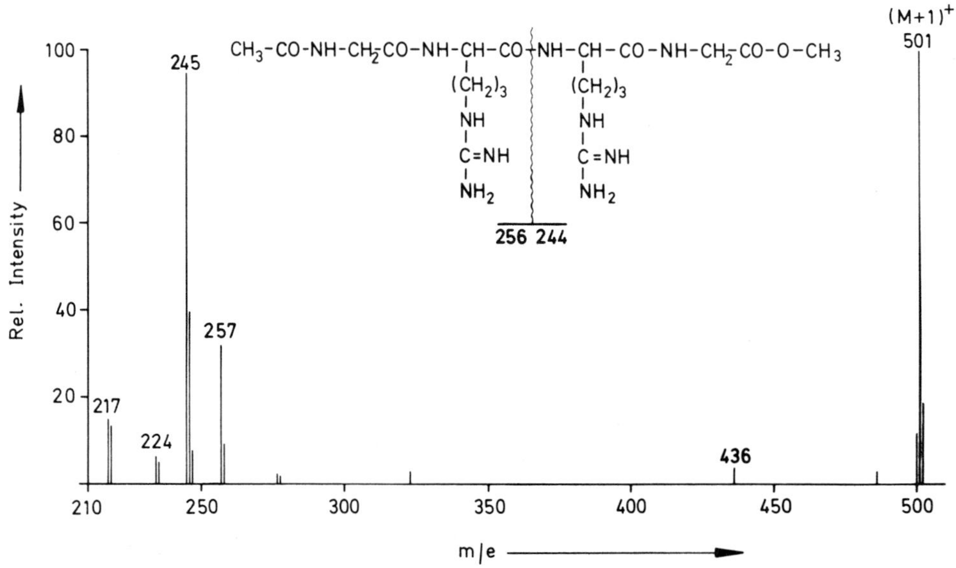

FIG. 40. FD mass spectrum of Ac-Gly-Arg-Arg-Gly-OMe [111].

$[M + H]^+$ ion at m/e 501 is the salient feature of the spectrum, because the peptide contains among other groups two unprotected arginines which greatly reduce the volatility. The direct bond rupture (with hydrogen transfer) in the middle of the molecule leads to two sequence-specific fragments. As already explained in Sections VI. A. 1. b, VI. B. 1. b, and VI. B. 1. d., one obtains structure information from the molecular ion and the complementary signals at m/e 256 and m/e 244 which are relatively easy to interpret. One must, however, make the qualifying statement that in this case the identity of the individual amino acids and their arrangement are obtained only from additional data of EI-MS. Hence, if the complete sequence is to be determined, a combination of EI and FD (or CI)-MS would probably have the greatest chance of success. But there exists a study suggesting that, with FD-MS only, sufficient fragmentation of the peptide skeleton was obtained to describe a complete sequencing of an unprotected tri- and nonapeptide [112]. On the other hand, the attempt to generate thermally- and field-induced fragments considerably above T* by higher emitter heating currents leads predominantly to pyrolytic products and not to sequence fragments.

4. Peptide Mixtures

An important analytical problem in biochemistry and medicine is the analysis of mixtures of peptides. On the one hand, nearly all living tissues appear to contain appreciable quantities of low-molecular-weight peptides. On the other hand, the most interesting group of biologically active large peptides and proteins are out of the range of direct mass spectrometric investigation. Therefore they are split chemically into smaller subunits such as amino acids or small peptides and are then amenable to mass spectrometry (as well as other analytical methods, e.g., by the Edman Sequenator). One essential problem, however, is that the free peptides cannot be introduced into the mass spectrometer as such, but have to be derivatized to gain sufficient volatility for vaporization. A number of different protecting groups and special techniques have been introduced for this purpose. But there are considerable difficulties when the mixtures which are obtained have a very complex composition and the material of interest is only obtainable in minute amounts. In order to explore the possibilities of FD-MS for this analytical problem, we investigated a model mixture of three underivatized dipeptides. The FD spectrum of this mixture of Gly-Ala, Gly-Val, and Gly-Leu is shown in Figure 41. The molecular ions and protonated molecules of all three peptides give prominent signals in the spectrum with the exception of Gly-Leu which gives only the $[M + H]^+$ ion and no $[M]^{+\cdot}$ ion. The process of water elimination from the protonated molecules is in general very pronounced.

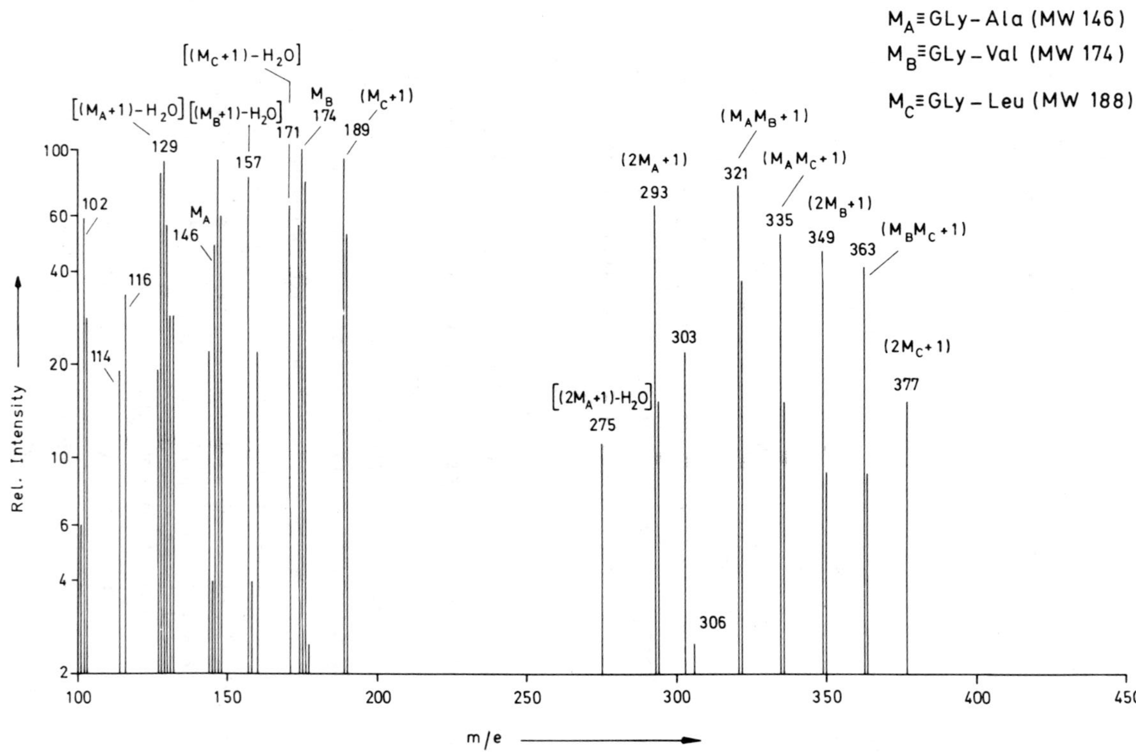

FIG. 41. FD mass spectrum of an equimolar (10^{-2} M) aqueous solution of three dipeptides, photographic detection on vacuum-evaporated AgBr plates [99].

This indicates that the time/temperature program which was employed was chosen for desorption above T* in order to induce thermal fragmentation. The emitter heating current was raised at a constant rate from 0 to 50 mA during the 11 min of exposure time. The photoplate as an integrating recording device collected all the ions generated during this desorption procedure. This also holds true for the peaks in the upper mass range (m/e 275 to m/e 377). At first glance this part of the spectrum is somewhat puzzling, but high-resolution FD-MS reveals that these signals contain very useful information for the identification of the different chemical species in the mixture. For all three free peptides the $[2M + 1]^+$ ions are easily found, and these cluster ions can be used to confirm the molecular weight of the individual peptides. In addition all mixed clusters of Gly-Ala, Gly-Val, and Gly-Leu are detected, so that in this first example of a peptide mixture the unambiguous assignment of the three dipeptides can be performed. Since the synthetic mixture of underivatized peptides gave such a meaningful and easily understandable mass spectrum, FD-MS appears to be a promising technique for analyzing peptide mixtures and thus might give an answer to the problems mentioned above.

5. *Lipids*

EI-MS has proved useful in the structure determination and identification of lipids. However, some of these compounds are not amenable to conventional mass spectrometric methods unless in the form of a derivative. Problems arise for the determination of the molecular weight, particularly if only small amounts of material are available. Evans et al. [113] showed that the FD mass spectra of triglycerides and phosphoglycerides enabled the molecular species to be readily identified. In addition natural mixtures of olive and safflower oil were studied and gave relatively high abundance of the molecular ions of the different

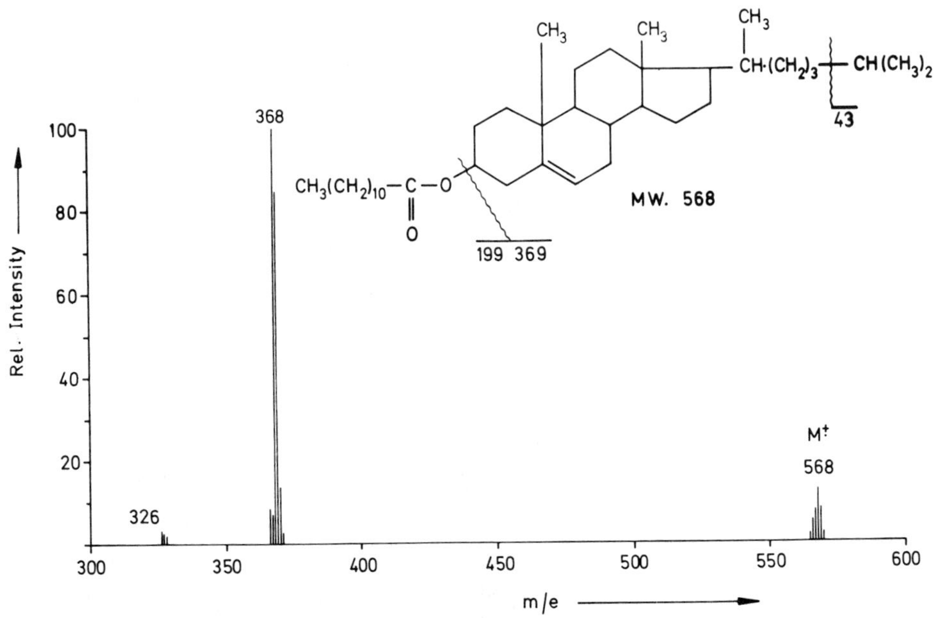

FIG. 42. FD mass spectrum of cholesteryllaurate [99]. Electric detection with a Varian MAT 731 instrument; solvent ethyl acetate; source temperature 80°C, 9-10 mA heating current.

compounds of the mixture. Qualitative analysis of a mixture of phosphatidylcholine from egg yolk by FD-MS reflected the composition of the different long-chain fatty acids. Wood and Lau [114] reported the analysis of intact phospholipids by FD-MS. Diacylglycerophosphorylcholines (lecithins) give prominent molecular ions. For 1,2-dipalmitoyl-glycero-3-phosphorylcholine a $[M]^{+\cdot}$ ion was obtained with 90% relative abundance. In Figure 42 the FD spectrum of cholesteryl-laurate is displayed. Here the molecular ion is only of 13% rel. int. The direct bond cleavage between the fatty acid moiety and the cholesteryl part leads to the base peak of the spectrum at m/e 368. Apparently this cleavage is accompanied by hydrogen transfer to the lauric acid. A minor process is the loss of C_3H_7 from the sidechain that gives a peak at m/e 326. In view of the importance of the substances described above in biochemistry and medicine, further application of FD-MS for qualitative *and* quantitative analysis is indicated, and a comprehensive survey on the application of FD-MS in biochemical analysis has recently been published [99].

C. Application in Medicine: Drugs and Drug Metabolites

1. Anticancer Drugs

For one of the authors (H.-R. S.) the FD investigations of these compounds constitute, since 1971, the focal point in testing the method [115,116]. A number of propane and butane sultames (antitumor effect against Ehrlich ascites tumors) have been measured with EI- and FD-MS and compared [37]. Five general aspects were found for the FD spectra: high intensity of molecular ions, one-step formation of the majority of fragments, elimination of neutral compounds (H_2O, NH_3, HCOOH) from the protonated molecule, formation of fragments by direct bond rupture, and the generation of doubly charged ions.

The publications on cyclophosphamide (A) (Endoxan, 2[bis-(2-chloroethylamino]tetrahydro-2H-1,3,2-oxazaphosphorine-2-oxide) and its metabolites [90, 117] go beyond model compounds toward the solution of "real" analytical problems with FD. It is shown that single compounds can be identified with highly resolved FD-MS not only in mixtures of primary compounds (as in Sec. VI. A. 2), but also in mixtures of primary compounds and their metabolites. An excerpt of an FD spectrum of a mixture of cyclophosphamide and three of its most important known metabolites (B, C, D) is seen in Figure 43. Although these compounds are spread over a large range of polarity (D was stabilized as cyclohexylammonium salt), the molecular ions of all components appear with high relative intensity. Also, in extracts of body fluids,

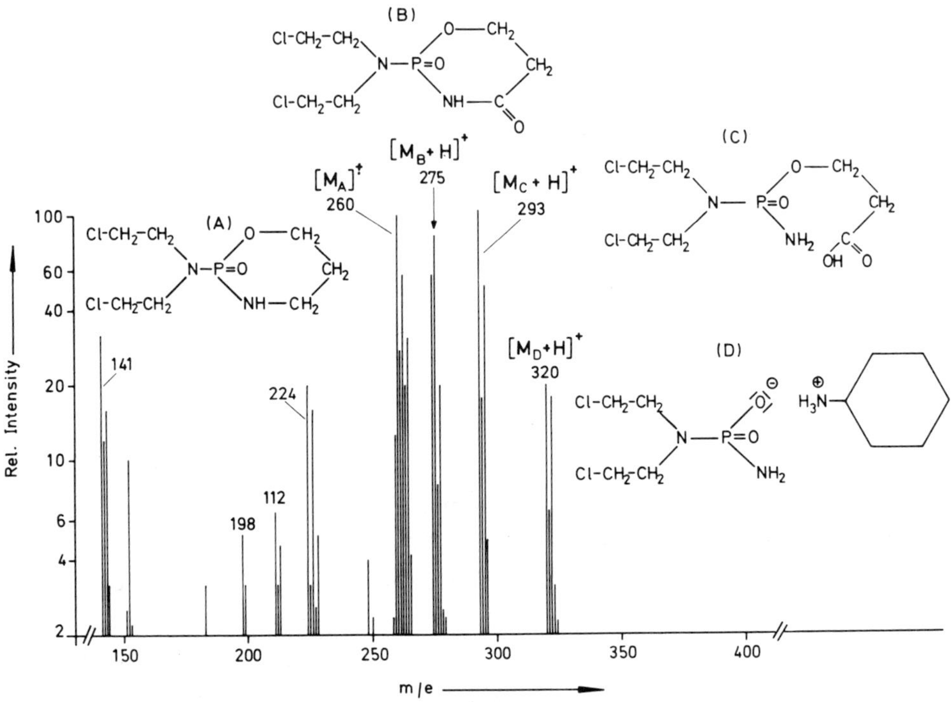

FIG. 43. FD mass spectrum of a solution/suspension of 50 μg of compound A (cyclophosphamide), B (4-ketocyclophosphamide), C (carboxyphosphamide), and D [N,N-bis-(2-chloroethyl)phosphorodiamidic acid cyclohexylammonium salt] in 10 μl acetone. Only a part of the spectrum between m/e 141 and m/e 400 is displayed (see Ref. 90).

it was possible to detect (subsequent to thin-layer chromatography) decomposition products of cyclophosphamide. However, due to impurities that were carried along, the FD spectra displayed many peaks and looked rather like those obtained with electron impact. Without additional information about the compound in question the interpretation of these spectra is very difficult [118].

In addition to oxidation products of drugs, conjugates are of particular interest. One of the usual reactions of alkylating agents in the body is conjugation with cystein or N-acetyl-

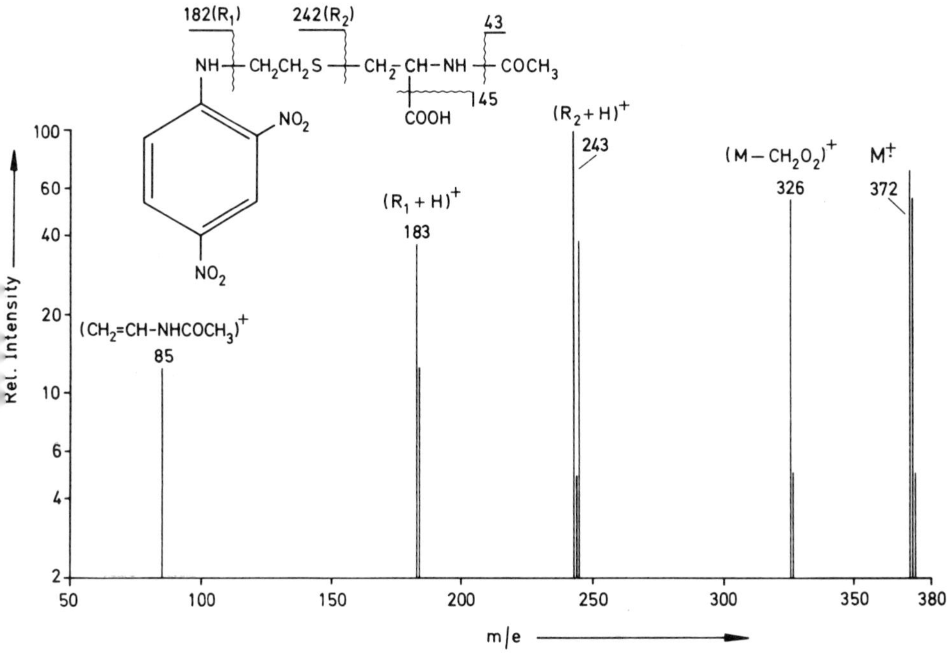

FIG. 44. High-resolution FD mass spectrum of S-[N-(2,4-dinitrophenyl)]mercapturic acid. The average error in the mass determination was ±3 millimass units [99].

cysteine. As an example for analyzing these conjugates *without* derivatization, the FD spectrum of S-[N-(2,4-dinitrophenyl]mercapturic acid is shown in Figure 44 [99]. This compound occurs as urinary metabolite of the anti-tumor agent 1-(1-arizidinyl)-2,4-dinitrobenzene in the rat [119]. The molecular ion (m/e 372.074) of the free acid is displayed with 70% rel. int. in the FD spectrum. Direct bond cleavage and protonation yield two structurally diagnostic ions at m/e 183.028 $[R_1 + H]^+$ and 243.031 $[R_2 + H]^+$. In addition the generally observed loss of formic acid in the FD spectra of free carboxylic acids produces a prominent ion at m/e 326.068. Fragments of smaller relative intensity are recorded at m/e 85.0053 $[C_4H_7NO]^+$ and 43.018

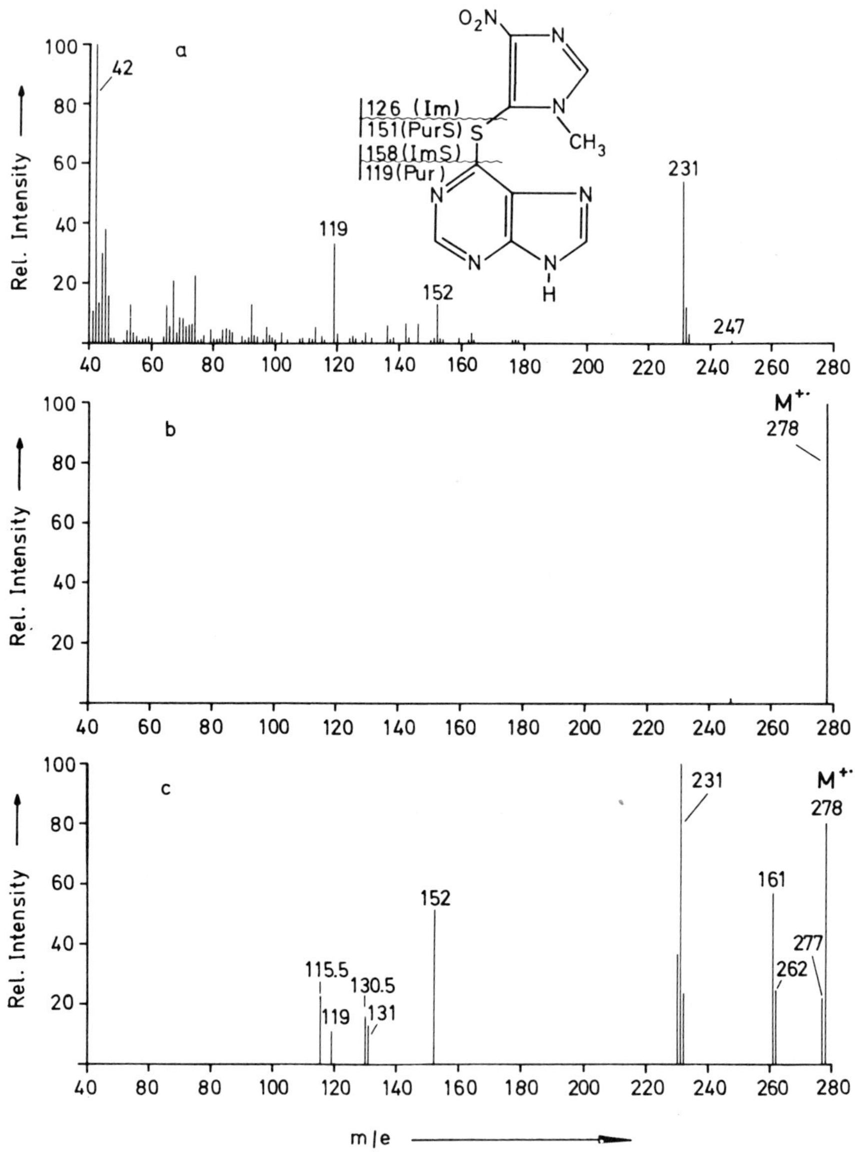

FIG. 45. Mass spectra of Azathioprine: (a) electron impact, using a direct probe at a temperature of 210°C; (b) FD, using 22 mA emitter heating current (= BAT); (c) FD, using 25 mA emitter heating current (> BAT) [36].

$[C_2H_3]^+$. Likewise the cysteine conjugate of the nitrobenzene derivative gave similar results with FD-MS. The obtained molecular ion intensities, however, are smaller (about 30% rel. int.) when the spectrum is recorded on a Varian CH4 instrument with electric detection [99].

2. Immunosuppressive Drugs

A widely applied drug of this group is Azathioprine (Imuran, [(1-methyl-4-nitroimidazolyl)thio]purine). The EI spectrum of this compound is shown in Figure 45a. The relative intensity of the molecular ion is below 1% [36]. At T* the FD spectrum exhibits only the $[M]^{+\cdot}$ peak (Fig. 45b). At heating currents greater than T* we get the FD spectrum displayed in Figure 45c. Without discussing in detail the fragments obtained, it becomes evident that ions in the higher-mass range of the EI-MS and FD-MS at 25 mA give similar fragments (e.g., m/e 231, 152, 119). Hence, obviously, the predominantly thermal decay of the Azathioprine molecule with EI is simulated by the thermally and field-induced fragmentation with FD. One can find the same elementary composition of ions for EI and FD by high-resolution mass measurements, without taking the formation mechanism of these ions into consideration. More important in this context is a metabolite of Azathioprine that is a glutathionyl conjugate. The EI-MS in Figure 46a is characterized predominantly by pyrolytic decay of the molecule. At T*, the FD-MS again gives only the $[M + H]^+$ ion (Fig. 46b). Above T* it is possible to induce fragmentation in a very narrow heating current interval (Fig. 46c).

In order to investigate the possibilities of field desorption in the detection of metabolites, a sample was isolated from the urine of a rat which had been given Azathioprine orally. After column chromatography and purification by paper chromatography a cysteinyl conjugate could be identified [36]. A more detailed survey of the main reasons for the usefullness of FD-

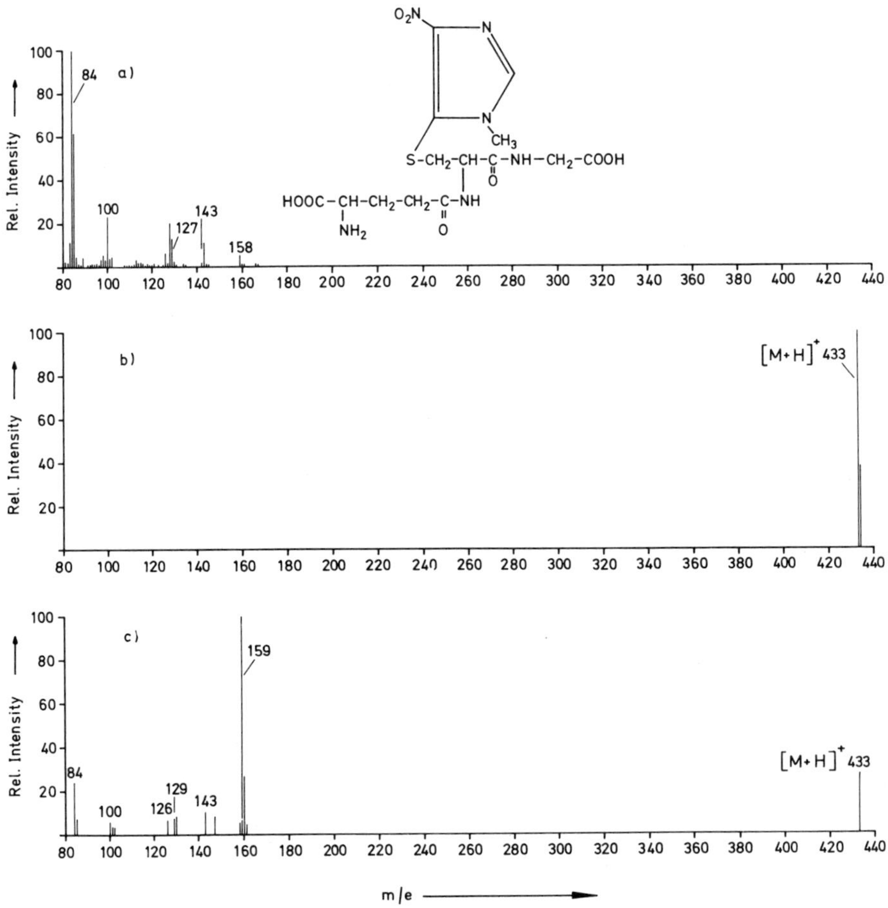

FIG. 46. Mass spectra of 1-methyl-5-nitro-(S-glutathionyl)imidazole; (a) electron impact, using a direct probe at a temperature of 195°C; (b) FD, using 24 mA emitter heating current (= BAT); (c) FD, using 27 mA emitter heating current (> BAT) [36].

MS in the study of biotransformation processes has been given in Refs. 99 and 116.

VII. ADVANTAGES AND DISADVANTAGES OF THE FD METHOD AS COMPARED TO OTHER IONIZATION METHODS

Let us compare the FD method with other means of ionization which can also be regarded as "soft," such as CI, FI, and EI at low electron energies. The FD method is very well suited for the qualitative analysis of complicated mixtures of substances, and it is adequate for quantitative analyses of single compounds and of mixtures. There is, however, one main problem. For each component of the mixture there exists a definite optimum temperature T* of the field anode, as presented in Section IV.D. Thus, at constant anode temperature, the FD ion current signals are certainly not proportional to the molar entities of the components of a mixture in solution. Even if one were to introduce a variable-temperature program for the field anode, it would hardly be possible to reproduce such a proportionality. The ion density of each single substance depends in a very complicated manner on the structure and the chemical properties of the organic microneedles on the emitter surface, on the concentrations of the mixture components, on the thickness of the sample layer deposited on the microneedles, on the solvent, and on other parameters (e.g., on how long the sample remains on the field anode during field desorption). From these reasons it becomes clear that the method of choice for quantitation with FD-MS is the use of stable isotope dilution.

The strength of the FD method doubtlessly lies in the area of qualitative analysis of thermally labile substances. However, it must be emphasized that, at the present level of development of the technique, the reproducibility is less than that of EI or CI. But, this aspect is generally of minor importance, because the FD method is used in cases where neither CI and FI nor EI can produce clearly identifiable molecular ions in the samples due to their nonvolatility. Under these conditions, the combination of FD with EI and/or CI is complementary.

A great advantage of the FD method is the fact that substances which are present only in picogram amounts do not need to be chemically treated to increase their volatility. The ability to analyze very small amounts of strongly polar underivatized substances is thus one of the crucial considerations in the application of the FD method.

The electron impact method has the advantage that numerous structure-specific molecule fragments are found in the mass spectrum. In the approximately 30 years during which the EI method has been used in a great number of laboratories over the entire world, extensive experimental results have been gathered which are documented in voluminous catalogs and in numerous empirical intensity rules for the fragmentation processes.

The method of chemical ionization introduced by Munson and Field [120] has been used for a few years in a relatively large number of laboratories, while until 1974 the FD method has been systematically tested and developed almost exclusively by the group at Bonn. Nonetheless at this stage, numerous specific application possibilities have already been delineated, some of which have been discussed in this article.

VIII. FUTURE DEVELOPMENTS

A. FD-Quadrupole Mass Spectrometry

FD-quadrupole mass spectrometry (FD-QMS) is a field that is still in the initial stage of development, which, however, could become of significance for special branches of FD mass spectrometry. In Section IV. C the importance of a rapid mass scan was pointed out for the case of electrical ion detection.

The QMS has specific properties and advantages for the FD technique at low resolution ($m/\Delta m < 500$) [121].

1. A mass scan rate of the order of 1000 mass units per second can be achieved.

2. The mass resolution and the intensity of the ion current measured at the ion detector is almost independent of the ion kinetic energy in the range of about 8 to 12 eV. (The energy of the ions entering the QMS should not be much higher than this figure.)
3. The product of transmission (T) and resolution (R) is approximately constant in the lower-mass range [122]. The transmission is defined as the ratio of the current measured at the first dynode of the secondary electron multiplier used for ion detection and the total ion current emitted by the field anode. The ratio T/R can be controlled electronically with the QMS. Therefore one can choose either a large-mass range at large transmission and low resolution (for fast inspection of the whole FD mass spectrum), or increase the resolution at the expense of the transmission, e.g., for more exact analysis of the molecular ion group.

The high scan rate of the QMS enables one to match the recording of the FD mass spectrum to the specific desorption characteristics of the substance to be analyzed. This is important for the study of fundamental phenomena of the FD process, such as desorption kinetics, and also for analytical applications of the FD method. The rapid detection of certain components of mixtures may be also mentioned, which originate from the chemical synthesis of compounds or with methods of extraction in biochemistry [123].

B. Pyrolysis Mass Spectrometry

Some of the most important biologically active compounds such as polysaccharides, polynucleotides, and polypeptides have very high molecular weights (>20,000). In the near future it cannot be expected that these macromolecules can be ionized without decomposition and detected in a routine mass spectrometer. Controlled thermal degradation and identification of the structural units by mass spectrometric methods holds out good prospects in two directions: (1) Rapid, accurate, and reproducible analysis of the pyrolysis products, e.g., with a combination of Curie-point pyrolyzer, quadrupole MS, and data system, leading to fingerprint spectra [124, 125]. The use of this system with the comparison of the spectra can provide unambiguous identification of known polymers. (2) Identification of ionized primary fragments in the mass spectrum provides indications of the composition and structure of known and unknown polymers. A recently developed technique, pyrolysis-field desorption (Py-FD-MS), is suitable for investigations on bipolymers [126, 127] and industrial polymerization products [128]. Py-FD-MS is the direct pyrolysis on the surface of a high-temperature-activated tungsten wire emitter. This method is necessary if one wants to obtain structural information about compounds with a molecular weight which considerably exceeds m/e 5000. The larger the weight of the detected ions, the more valuable the information obtained. The observation of these large primary products is favored by four important conditions:

Minimal difference in time and place between pyrolysis and ionization
Minimal sample quantities
Minimal residence time of the free ions in the pyrolysis zone
Minimal excess energy transferred by ionization.

Py-FD-MS fulfills all these requirements to a high degree: First, pyrolysis and ionization coincide in place and time; second, the sample quantity can be varied between 10^{-6} and 10^{-9} g; third, the residence time of the free ion in the pyrolysis zone is of the order of 10^{-11} sec; fourth and finally, the average excess energy does not exceed a few tenths of an electron volt. The first objects of our investigations were biopolymers such as DNA [127], RNA [64], and polysaccharides [126] where high resolution (∿20,000, 10% valley definition) enables the elemental composition of large structure units to be determined. In Figure 47 the first example for a Py-FD spectrum of a technical polymer, polyethylene, is given. This spectrum is especially interesting for two reasons: First, the pattern of the CH_2

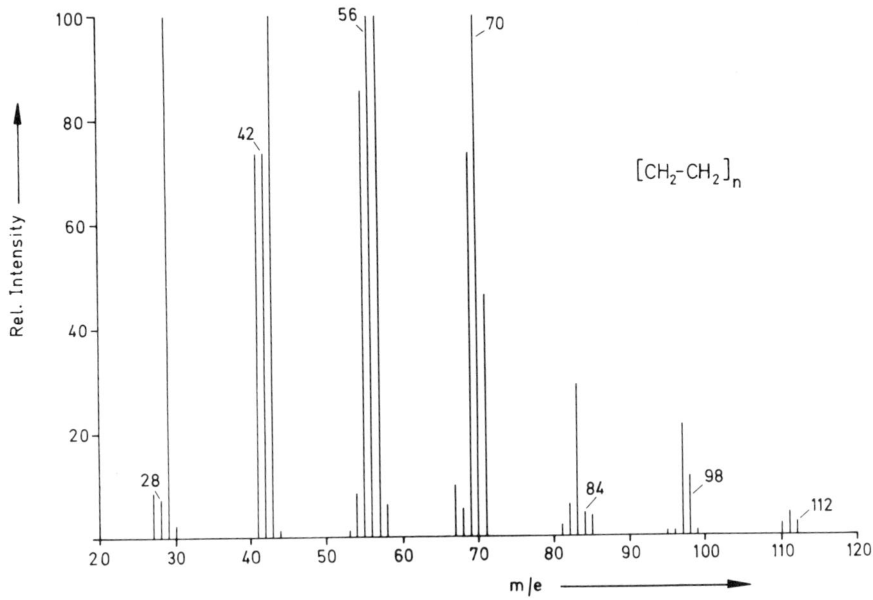

FIG. 47. Pyrolysis-field-desorption mass spectrum of a fine suspension of polyethylene, automatically regulated time/temperature program for desorption and pyrolysis, sample consumption $\sim 10^{-8}$ g [128].

groups (14 mass units) reflects the stability of the ions produced and the structure of the polymer. It is still an open question whether or not branching and other structural details of the polymer chain can be directly correlated with the ion species obtained. Investigations of the other technical products like polyacrylonitrile and polyacrylamide showed, however, that with a controlled time/temperature program for the emitter heating current and photographic detection, significant FD spectra can be obtained. Second, when the polyethylene was ground to a fine powder at a very low temperature (liquid N_2) and applied to the emitter from a suspension in acetone, it becomes evident that not only solutions but also suspensions can be analyzed with FD-MS.

It must be emphasized that this promising pyrolysis-field desorption combination can be applied only when two fundamental prerequisites are satisfied. In the first place, very high demands are made on the resolving power of the mass spectrometer. In the absence of a high-resolution MS, the assignment of the elemental composition of the peaks runs into increasing difficulty with increasing mass number. Secondly, the extremely complicated spectra of multicomponent mixtures encountered can be evaluated only if electronic data processing is available.

C. Liquid Chromatography and Field Desorption

The coupling of a gas chromatograph to the mass spectrometer is regarded as a very successful and useful combination. However, compounds are amenable to this system only if they themselves are sufficiently volatile or if this volatility can be produced by chemical pretreatment, i.e., by the preparation of derivatives. In view of the rapid progress in high-pressure liquid chromatography and in field desorption, both of which avoid the necessity of vaporizing the analytical sample, a combination of the two methods seems highly desirable and has been accomplished.
The first experiments in this direction were made by a two-step technique [129]: The eluate from the liquid chromatograph was collected, concentrated on the emitter, and analyzed by FD spectrometry. In this sequence the FD mass spectrometer serves as a (second) molecule-specific detector. The use of column liquid chromatography in combination with field desorption mass spectrometry is exemplified by the separation and identification of the components of a steroid mixture extracted from rat serum [129]. The liquid chromatogram of this mixture is shown in Figure 48 (the experimental conditions are described in Ref. 129). The effluent of the high-pressure chromatograph was collected and transferred on the surface of the field desorption emitter. The FD spectrum of prednisolone (Fig. 49) shows the molecular

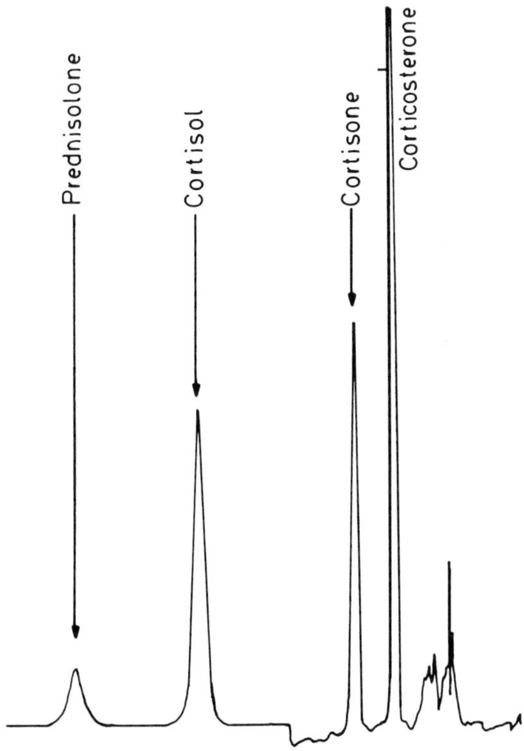

FIG. 48. Column high-pressure liquid chromatogram of a steroid mixture from rat serum [129].

ions as the base peak of the spectrum. Only one fragment is observed due to a common cleavage in the side chain between C(20) and C(21). Progesterone and corticosterone yield similar FD spectra.

The next stage in future investigations will involve an attempt to identify an unknown chemical species hidden in the peak of a liquid chromatogram, and later perhaps the sophisticated technical problem of an in-line coupling will be attacked. It remains to be seen whether the great efforts devoted to the inline technique [130] will lead to a solution of the

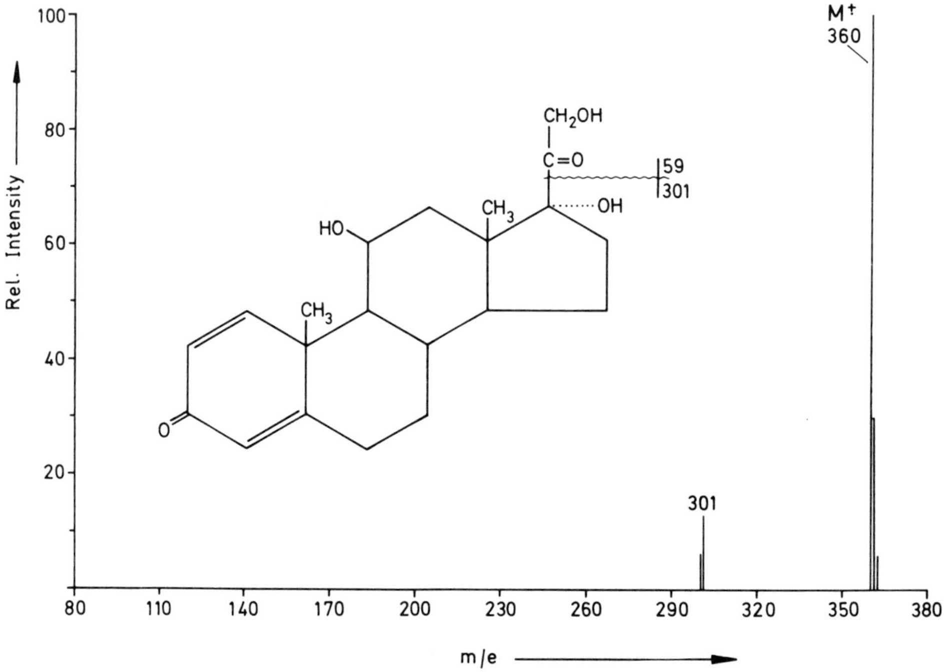

FIG. 49. High resolution FD-MS of prednisolone [129].

technical problems, especially for highly polar substances, and whether the analytical results of this direct coupling will justify the high cost of the apparatus.

D. Inorganic Chemistry and Field Desorption

This chapter has dealt almost exclusively with organic substances. Now we would like to report some results from "the other half of the chemistry world," namely, inorganic chemistry. The first complete and consistent FD spectra in this respect were obtained from alkali halides [97]. In general, extraordinarily high relative intensities for the cations (C) were recorded. Particularly, $[Li]^+$, $[Na]^+$, and $[K]^+$ ions gave ion current intensities that were by orders of magnitude greater than those for all

TABLE 3

Applications of Field Desorption Mass Spectrometry in Inorganic Chemistry[a]

Salt	The three most intense ions in the FD spectra (except the cations)
LiF	$(LiF)Li^+$, $(LiF)_2Li^+$, $(LiF)_3Li^+$
LiCl	$(LiCl)Li^+$, $(LiCl)_2Li^+$, $LiCl^+$
LiJ	$(LiJ)Li^+$, J^+, $(LiJ)_2Li^+$
KF	$(KF)K^+$, $(KF)_2K^+$
KCl	$(KCl)K^+$, $(KCl)_2K^+$, KCl^+
KBr	$(KBr)K^+$, Br^+
KJ	$(KJ)K^+$, $(KJ)_2K^+$, J^+
NaCl	$(NaCl)Na^+$, $(NaCl)_2Na^+$, $NaCl^+$
$NaNO_3$	$(NaNO_3)Na^+$, $(NaNO_3)_2Na^+$, $(NaNO_3)_3Na^+$
Na_2SO_4	$(Na_2SO_4)Na^+$
$NaClO_3$	$(NaClO_3)Na^+$, $(NaClO_3)_2Na^+$, $(NaCl)Na^+$
$Na(CH_3CO_2)$	$[Na(CH_3CO_2)]Na^+$, $[Na(CH_3CO_2)]_2Na^+$
NaN_3	$(NaN_3)Na^+$
AgCl	$AgCl^{+\bullet}$, $(AgCl)Ag^+$
$AgNO_3$	$(AgNO_3)Ag^+$, $(AgNO_3)_2Ag^+$
$CaCl_2$	$CaCl^+$, $Ca^{+\bullet}CaCl_2^{+\bullet}$

[a]Ions and salt clusters in the FD mass spectra of some inorganic salts.

other types of ions ever observed in FD-MS. The high intensities observed are explained by field assisted thermoionization of the Saha-Langmuir type. In addition to $[C]^+$ ions, the energetically and kinetically favored process of cluster formation generates ions from which the molecular structure can be easily derived [84]. In Table 3 the three most abundant ions (except for the metal cations) in the FD spectra of some representative inorganic salts with simple and complex anions (A) are listed [87]. Salts with large organic cations such as quaternary ammonium salts [83, 84, 85, 87], phosphonium [84,86], arsonium, sulfonium, sulfoxonium, and oxonium salts [132] display high cation intensities. Note, that the $[C_{n+1} A_n]^+$ ion with n > 1 were not found in all cases. The FD spectra of five tricarbonyl (π-hydrocarbon) metal salts and cyclopentadienyl (arene) metal salts have been reported by Games et al. [133]. The same group also published the FD spectra of some phthalocyanines which normally gave the molecular ions as the base peak of the spectrum [134].

phthalocyanines which normally gave the molecular ions as the base peak of the spectrum [134].

From the analytical point of view, the knowledge of the characteristics of the ion formation in the field desorption of salts is essential for the interpretation of the FD spectra. Even mixtures of inorganic salts present in naturally occurring aerosols, which have been sampled in a newly developed method [135], were qualitatively identified by FD-MS [136].

Thus the method offers its potential to trace analysis of metal ions, detection and identification of inorganic salts, and molecular weight determination plus structural information of complex, involatile inorganic substances. Thereby FD-MS considerably increases the scope of mass spectrometric investigations.

The more recent investigations with FD-MS include the influence of inorganic salts on the obtained spectra. Elimination of surplus salt cations by the formation of macrocyclic ligand complexes [131] and addition of specific metal cations such as $[Li]^+$ for cationization of large organic molecules have been reported [143-145]. The extraordinary high sensitivity of the FD method for alkali ions led to a number of quantitative applications. Cesium was determined in spectrograde solvents, body fluids, environmental samples [146] and in heart cells [147]. By use of the stable isotope dilution technique quantitative determinations of high sensitivity and accuracy can be accomplished without prior treatment of the original samples as has been shown for thallium in brain tissue [148] and lithium in human body fluids [149]. Using a multichannel analyzer for accumulation of the FD spectra recorded by repetitive magnetic scan enables direct isotope determination of ions of inorganic [150, 151] and organic substances [152]. In order to demonstrate the capabilities of the FD method for quantitation, the methodology was discussed and examples were presented [153]. A comprehensive and complete survey of the present state of art of quantitative FD-MS has been given [154].

Further progress of the FD method was described for the combination with high pressure liquid chromatography [155-157] and the qualitative analysis of substances of high molecular weight. These latter studies include new mass references above m/e 1000 [158,159] and molecular weight determination and structural elucidation of peptide antibiotics [160], steroid- and triterpene saponins [161], and physiologically active pennogenin- and hederagenin glycosides [162]. Owing to the improvements in FD-MS using direct heating of the emitter wire by a LASER [163] the complete mass spectrum of vitamin B_{12} could be obtained [164].

ACKNOWLEDGMENTS

We thank the Deutsche Forschungsgemeinschaft, the Ministerium für Wissenschaft und Forschung des Landes Nordrhein-Westfalen, and the Fonds der Chemischen Industrie for generous financial support of our work.

REFERENCES

1. H. D. Beckey, *Principles of Field Ionization and Field Desorption Mass Spectrometry*, Pergamon, Oxford, 1977.
2. J. Block, *CRC Critical Rev. Solid State Sci.*, pp. 133-156 (1976).
3. A. J. B. Robertson, *MTP International Review of Science, Mass Spectrometry* (A. Maccol, Ed.), Physical Chemistry, Series One, Vol. 5, pp. 103-131.
4. H. D. Beckey, K. Levsen, F. W. Röllgen and H.-R. Schulten, *Surface Sci., 70*, 325 (1978).
5. K. Levsen, *Fundamental Aspects of Organic Mass Spectrometry, Progress in Mass Spectrometry*, (H. Budzikiewicz, Ed.), *Vol. 4*, pp. 53-57 and pp. 186-196 (1978).
6. P. J. Derrick and A. L. Burlingame, *Acc. Chem. Res., 7*, 328 (1974).

7. G. W. A. Milne and M. J. Lacey, *Crit. Rev. Anal. Chem., 3,* 45 (1974).

8. R. Gomer, *Field Emission and Field Ionization*, Harvard University Press, Cambridge, 1961.

9. E. W. Müller and T. T. Tsong, *Field Ion Microscopy*, Elsevier, New York, 1969.

10. R. Gomer, *J. Chem. Phys., 31*, 341 (1955).

11. E. W. Müller, J. A. Panitz, and S. B. McLane, *Rev. Sci. Instrum., 39*, 83 (1968).

12. E. W. Müller, *Ergeb. Exakten Naturwiss., 27*, 290 (1953).

13. H. G. Metzinger and H. D. Beckey, *Z. Physik. Chem., 52*, 27 (1967).

14. M. J. Southon and D. Brandon, *Philos. Mag., 8*, 579 (1973).

15. F. Speier, H. J. Heinen, and H. D. Beckey, *Messtechnik, 80*, 147 (1972).

16. F. W. Röllgen and H. D. Beckey, *Surf. Sci., 27*, 321 (1971).

17. F. W. Röllgen and H. D. Beckey, *Ber. Bunsenges. Phys. Chem., 75*, 988 (1971).

18. E. W. Müller, *Phys. Rev., 102*, 618 (1962).

19. M. G. Inghram and R. Gomer, *J. Chem. Phys., 22*, 1279 (1954); *Z. Naturforsch., 10a*, 863 (1955).

20. R. Ajeian, A. Maas, U. Nitschke, and H. D. Beckey, *Appl. Phys., 6*, 111 (1975).

21. H. D. Beckey, *J. Phys. E.,* in press.

22. E. W. Müller and K. Bahadur, *Phys. Rev., 102*, 624 (1956).

23. M. Anbar and W. H. Aberth, *Anal. Chem., 46*, 59A (1974).

24. H. D. Beckey, S. Bloching, M. D. Migahed, E. Ochterbeck, and H.-R. Schulten, *Int. J. Mass Spectrom. Ion Phys., 8*, 169 (1972).

25. H.-R. Schulten and H. D. Beckey, *Org. Mass Spectrom., 6*, 885 (1972).

26. H. D. Beckey, E. Hilt, and H.-R. Schulten, *J. Phys. E., 6*, 1043 (1973).

27. D. F. Barofsky and E. Barofsky, *Int. J. Mass Spectrom. Ion Phys., 14*, 3 (1974).

28. H. U. Winkler, W. Neumann, B. Fassbender, E. Hilt, and H. D. Beckey, *Messtechnik, 82*, 119 (1974).

29. H. D. Beckey, *Int. J. Mass Spectrom. Ion Phys., 2*, 500 (1969).

30. K. H. Maurer, U. Rapp, K. Chan, and W. Sadée, in *Advances in Mass Spectrometry in Biochemistry and Mecicine*, Vol. I (A. Frigerio and N. Castagnoli, Eds.), Spectrum Publications, New York, 1975.

31. H. D. Beckey, A. Heinrichs, and H. U. Winkler, *Int. J. Mass Spectrom. Ion Phys., 3*, App. 9-11 (1970).

32. H. D. Beckey and H.-R. Schulten, *Angew. Chem., 87*, 425 (1975); *Angew. Chem. Int. Ed. Eng., 14*, 403 (1975).

33. K. L. Olson, J. C. Cook, Jr., and K. L. Rinehard, Jr., *Biomed. Mass Spectrom., 1*, 358 (1974).

34. D. Kümmler and H.-R. Schulten, *Org. Mass Spectrom., 10*, 813 (1975).

35. H. U. Winkler and H. D. Beckey, *Org. Mass Spectrom., 7*, 1007 (1973).

36. D. A. Brent, P. de Miranda, and H.-R. Schulten, *J. Pharm. Sci., 63*, 1370 (1974).

37. H.-R. Schulten and H. D. Beckey, *Org. Mass Spectrom., 7*, 861 (1973); H.-R. Schulten and N. M. N. Nibbering, *Biomed. Mass Spectrom., 4*, 55 (1977).

38. H.-R. Schulten, H. D. Beckey, G. Eckhardt and S. H. Doss, *Tetrahedron, 29*, 3861 (1973).

39. F. W. Röllgen and H.-R. Schulten, *Org. Mass Spectrom., 10*, 660 (1975).

40. H.-R. Schulten and H. D. Beckey, *Org. Mass Spectrom., 9*, 1154 (1974).

41. H. D. Beckey, *Angew. Chem., 81*, 662 (1969); *Angew. Chem. Int. Ed. Eng., 8*, 623 (1969).

42. H. Krone and H. D. Beckey, *Org. Mass Spectrom., 5*, 983 (1971).

43. K. Heyns, H. F. Grützmacher, H. Scharmann, and D. Müller, *Fortschr. Chem. Forsch., 5*, 448 (1966).

44. K. Heyns and D. Müller, *Tetrahedron, 21*, 55 (1965).

45. J. N. Damico, R. P. Barron, and J. A. Sphon, *Int. J. Mass Spectrom. Ion Phys., 2*, 161 (1969).

46. H.-R. Schulten and H. D. Beckey, *J. Agr. Food Chem., 21*, 372 (1973).

47. P. Brown, G. R. Pettit, and R. K. Robins, *Org. Mass Spectrom., 2*, 521 (1969).

48. P. Brown and G. R. Pettit, *Org. Mass Spectrom., 3*, 67 (1970).

49. P. Brown, F. Brüschweiler, R. G. Pettit, and T. Reichstein, *J. Amer. Chem. Soc., 92*, 4470 (1970).

50. P. Brown, F. Brüschweiler, R. G. Pettit, and T. Reichstein, *Org. Mass Spectrom.*, *5*, 573 (1971).

51. C. Fenselau, S. Y. Wang, and P. Brown, *Tetrahedron*, *26*, 5923 (1970).

52. P. Brown and C. Fenselau, *Org. Mass Spectrom.*, *7*, 305 (1973).

53. E. M. Chait, T. W. Channon, W. O. Perry, G. E. Van Lear, and F. W. McLafferty, *Int. J. Mass Spectrom. Ion Phys.*, *2*, 141 (1969).

54. I. V. Goldenfeld et al., *Theoret. Eksp. Khim.*, *6*, 548, 830, (1970); *7*, 370, 550, 885 (1971); *8*, 770, 777 (1971).

55. H. M. Fales, G. W. A Milne, H. U. Winkler, H. D. Beckey, J. N. Damico, and R. Barron, *Anal. Chem.*, *47*, 207 (1975).

56. N. J. Haskins, D. E. Games, and K. T. Taylor, *Biomed. Mass Spectrom.*, *1*, 423 (1974).

57. H. D. Beckey and G. Wagner, *Z. Anal. Chem.*, *197*, 58 (1963).

58. J. N. Damico and R. P. Barron, *Anal. Chem.*, *43*, 17 (1971).

59. H. D. R. Schüddemage and D. O. Hummel, *Advances in Mass Spectrometry*, Vol. 4 (E. Kendrick, Ed.), The Institute of Petroleum, London, 1968, p. 857.

60. J. B. Forehand and W. F. Kuhn, *Anal. Chem.*, *42*, 1839 (1970).

61. H.-R. Schulten, H. D. Beckey, H. L. C. Meuzelaar, and A. J. H. Boerboom, *Anal. Chem.*, *45*, 191 (1973).

62. D. E. Games, A. H. Jackson, and D. S. Millington, *Tetrahedron Lett.*, 3063 (1973).

63. D. E. Games, A. H. Jackson, D. S. Millington, and M. Rossiter, *Biomed. Mass Spectrom.*, *1*, 5 (1974).

64. M. A. Posthumus, N. M. M. Nibbering, A. J. H. Boerboom, and H.-R. Schulten, *Biomed. Mass Spectrom.*, *1*, 352 (1974).

65. K. Levsen and H.-R. Schulten, *Biomed. Mass Spectrom.*, *3*, 137 (1976).

66. J. N. Damico, R. P. Barron, and J. M. Ruth, *Org. Mass Spectrom.*, *1*, 331 (1968).

67. F. J. Biros, R. C. Dougherty, and J. Dalton, *Org. Mass Spectrom.*, *6*, 1161 (1972).

68. H.-R. Schulten, Research Conference on Pesticide Metabolites-Enzymology, Isolation and Characterization, Fargo, North Dakota, June 1972.

69. H.-R. Schulten, in *Conference Proceedings No. 27, Fourth Annual Symposium on Recent Advances in the Analytical Chemistry of Pollutants*, Basle, June 17-19, 1974.

70. H.-R. Schulten, H. Prinz, H. D. Beckey, W. Tomberg, W. Klein, and F. Korte, *Chemosphere, 2*, 23 (1973).

71. S. Uk, C. M. Himel, and T. F. Dirks, *Bull. Environ. Contam. Toxicol., 8*, 97 (1972).

72. H.-R. Schulten, *Research Conference on New Methodology in Ecological Chemistry*, Susono City (Japan), Nov. 1973; see also F. Coulston, F. Korte, and M. Goto, *New Methods in Environmental Chemistry and Toxicology*, International Academic Printing Co., Tokyo, 1973, p. 31ff.

73. J. A. Sphon and J. N. Damico, *Org. Mass Spectrom., 3*, 51 (1970).

74. H.-R. Schulten and N. M. M. Nibbering, *Biomed. Mass Spectrom., 4*, 55 (1977).

75. F. W. Röllgen and H. D. Beckey, *Surf. Sci., 23*, 69 (1970).

76. F. W. Röllgen and H. D. Beckey, *Ber. Bunsenges. Phys. Chem., 75*, 988 (1971).

77. F. W. Röllgen and H. D. Beckey, *Adv. Mass Spectrom., 6*, 499 (1974).

78. D. E. Games, M. P. Games, A. H. Jackson, A. H. Olavesen, M. Rossiter, and P. J. Winterburn, *Tetrahedron Lett.*, 2377 (1974).

79. D. E. Games, A. H. Jackson, and M. Rossiter, unpublished.

80. H.-R. Schulten and H. D. Beckey, *Adv. Mass Spectrom., 6*, 499 (1974).

81. J. F. Abramson, *Anal. Chem., 44*, 28A (1972).

82. H.-R. Schulten, *J. Agr. Food. Chem., 24*, 743 (1976).

83. H.-R. Schulten and D. E. Games, *Biomed. Mass Spectrom., 1*, 120 (1974).

84. H.-R. Schulten and F. W. Röllgen, *Org. Mass Spectrom., 10*, 649 (1975).

85. D. A. Brent, D. J. Rouse, M. C. Sammons and M. M. Bursey, *Tetrahedron Lett., 42*, 4127 (1973).

86. G. W. Wood, J. M. McIntosh, and P.-Y. Lau, *J. Org. Chem., 40*, 636 (1975).

87. H.-R. Schulten and F. W. Röllgen, *Angew. Chem., 87*, 544 (1975); *Angew. Chem. Int. Ed., Eng., 14*, 561 (1975).

88. N. Evans, D. E. Games, M. J. E. Hewlins, J. F. J. Hughes, A. H. Jackson, J. R. Jackson, M. N. Khan, S. A. Matlin, R. Rossiter, R. G. Saxton, H. W. Swaine, and K. T. Taylor, in *Advances in Mass Spectrometry in Biochemistry and Medicine*,

Vol. I (A. Frigerio and N. Castagnoli, Eds.), Spectrum Publications, New York, 1975.

89. K. L. Rinehart, Jr., J. C. Cook, Jr., K. H. Maurer, and U. Rapp, *J. Antibiot., 27*, 1 (1974).

90. H.-R. Schulten, *Biomed. Mass Spectrom., 1*, 223 (1974).

91. H. Adlercreutz, B. Soltmann, and M. J. Tikkanen, *J. Steroid Biochem., 5*, 163 (1974).

92. H. D. Beckey and H.-R. Schulten, *Z. Anal. Chem., 273*, 345 (1975).

93. K. Nakanishi, D. A. Schooley, K. J. Judy, B. J. Bergot, M. S. Hall, J. B. Siddal, O. L. Chapman, J. A. Klun, K. C. Mattes, and M. Barry, in *New Methods in Environmental Chemistry and Toxicology* (F. Coulston, F. Korte, and M. Goto, Eds.), International Academic Printing Co., Tokyo, 1973, pp. 149-168.

94. H. Panzel and K. Ballschmiter, *Z. Anal. Chem., 271*, 182 (1974).

95. M. Suzuki, Y. Yamato, and M. Koga, *Biomed. Mass Spectrom., 5*, 518 (1978); H.-R. Schulten and I. Stöber, *Z. Anal. Chem., 293*, 370 (1978) and literature cited therein.

96. H.-R. Schulten and D. Kümmler, *Z. Anal. Chem., 278*, 13 (1976).

97. H.-R. Schulten and H. D. Beckey, in *Conference Proceedings No. 7, Twenty-Second Annual Conference on Mass Spectrometry and Allied Topics*, Philadelphia, Pennsylvania, May 19-24, 1974.

98. R. Large and H. Knof, *Chem. Commun.*, 935 (1974).

99. H.-R. Schulten, in *Methods of Biochemical Analysis*, Vol. 24, 313-448 (D. Glick, Ed.), Wiley, New York, 1977.

100. W. T. Shier, K. L. Rienhard, Jr., and G. Gottlieb, *Proc. Nat. Acad. Sci., USA, 63*, 198 (1969).

101. W. D. Lehmann, H.-R. Schulten, and H. D. Beckey, *Org. Mass Spectrom., 7*, 1103 (1973).

102. S. Caccamese, G. Montaudo, and M. Przybylski, *Org. Mass. Spectrom., 9*, 1114 (1974).

103. H.-R. Schulten and H. D. Beckey, in *Conference Proceedings Bl, Twenty-Third Annual Conference on Mass Spectrometry and Allied Topics*, Houston, Texas, May 25-30, 1975.

104. J. Moor and E. S. Waight, *Org. Mass Spectrom., 9*, 903 (1974).

105. M. M. Bursey and M. C. Sammons, *Carbohydr. Res., 37*, 355 (1974).

106. A. Bloch, "Drug Design," in *Medicinal Chemistry*, Vol. IV (E. J. Ariens, Ed.), Academic Press, New York, 1973, p. 286.

107. M. C. Sammons, M. M. Bursey, and D. A. Brent, *Biomed. Mass Spectrom., 1*, 169 (1974).

108. H.-R. Schulten, H. D. Beckey, E. M. Bessel, A. B. Foster, M. Jarman, and J. H. Westwood, *Chem. Commun.*, 416 (1973).

109. H. U. Winkler and H. D. Beckey, *Org. Mass Spectrom., 6*, 655 (1972).

110. H. D. Beckey, H. U. Winkler, K. H. Maurer, and G. Hoffmann, *Adv. Mass Spectrom., 5*, 626 (1971).

111. H. U. Winkler and H. D. Beckey, *Biochem. Biophys. Res. Commun., 46*, 391 (1972).

112. S. Asante-Poku, G. W. Wood, and D. E. Schmidt, Jr., *Biomed. Mass Spectrom., 2*, 121 (1975).

113. N. Evans, D. E. Games, J. L. Harwood, and A. H. Jackson, *Biochem. Soc. Trans., 2*, 1091 (1974).

114. G. W. Wood and P.-Y. Lau, *Biomed. Mass Spectrom., 1*, 154 (1974).

115. H. D. Beckey, H.-R. Schulten, and H. U. Winkler, *Proceedings of the 18th Field Emission Symposium*, Eindhoven, Netherlands, Aug. 1971, p. 81ff.

116. H.-R. Schulten and H. D. Beckey, *Twentieth Annual Conference on Mass Spectrometry and Allied Topics*, Dallas, Texas, June 4-9, 1972.

117. H.-R. Schulten, *Cancer Chemother. Rep., 60*, 501 (1976); M. Przybylski, H. Ringsdorf, U. Lenssen, G. Peter, G. Voelker, T. Wagner, and H. J. Hohorst, *Biomed. Mass Spectrom., 4*, 209 (1977); W. D. Lehmann and H.-R. Schulten, *Z. Anal. Chem., 290*, 121 (1978).

118. H.-R. Schulten, in *Advances in Biochemistry and Medicine*, Vol. I (A. Frigerio and N. Castagnoli, Eds.), Spectrum Publications, New York, 1975.

119. T. A. Connors, J. A. Hickman, M. Jarman, D. H. Melzack, and W. C. J. Ross, *Biochem. Pharmacol., 24*, 1665 (1975).

120. M. S. B. Munson and F. H. Field, *J. Amer. Chem. Soc., 88*, 2621 (1966).

121. H. J. Heinen, Ch. Hötzel, and H. D. Beckey, *Int. J. Spectrom. Ion Phys., 13*, 55 (1974).

122. W. Paul, H. P. Reinhard, and U. v. Zahn, *Z. Phys., 152*, 587 (1973).

123. H. Gierlich, H. J. Heinen, and H. D. Beckey, *Biomed. Mass Spectrom.*, *2*, 31 (1975).

124. H. L. C. Meuzelaar and P. G. Kistemaker, *Anal. Chem.*, *45*, 587 (1973).

125. H. L. C. Meuzelaar, M. A. Posthumus, P. G. Kistemaker, and J. Kistemaker, *Anal. Chem.*, *45*, 1546 (1973).

126. H.-R. Schulten, in *New Approaches to the Identification of Microorganisms* (C. G. Heden and T. Illeni, Eds.), Wiley, New York, 1974, pp. 117ff.

127. H.-R. Schulten, H. D. Beckey, A. J. H. Boerboom, and H. L. C. Meuzelaar, *Anal. Chem.*, *45*, 2358 (1973).

128. H.-R. Schulten and H. D. Beckey, in *Conference Proceedings No. H7, Twenty-First Annual Conference on Mass Spectrometry and Allied Topics*, San Francisco, May 20-25, 1973.

129. H.-R. Schulten and H. D. Beckey, *J. Chromatogr.*, *83*, 315 (1973).

130. P. Arpino, M. A. Baldwin, and F. W. McLafferty, *Biomed. Mass Spectrom.*, *1*, 80 (1974), and literature cited therein.

131. G. W. Wood, P.-Y. Lau, and N. Mak, *Biomed. Mass Spectrom.*, *1*, 425 (1974).

132. M. C. Sammons, M. M. Bursey, and C. K. White, *Anal. Chem.*, *47*, 1165 (1975).

133. D. E. Games, A. H. Jackson, L. A. P. Kane-Maguire, and K. Taylor, *J. Organometal. Chem.*, *88*, 345 (1975).

134. D. E. Games, A. H. Jackson, and K. T. Taylor, *Org. Mass Spectrom.*, *9*, 1245 (1974).

135. H.-R. Schulten and U. Schurath, *J. Phys. Chem.*, *79*, 51 (1975).

136. H.-R. Schulten and U. Schurath, *Atmos. Env.*, *9*, 1107 (1975).

137. J. F. Holland, B. Soltmann, and C. C. Sweeley, *Biomed. Mass Spectrom.*, *3*, 340 (1976).

138. U. Giessmann, H. J. Heinen, and F. W. Röllgen, *Org. Mass Spectrom.*, submitted for publication.

139. H. U. Winkler and B. Linden, *Org. Mass Spectrom.*, *11*, 327 (1976).

140. F. W. Röllgen, H. J. Heinen, and U. Giessmann, *Naturwiss.*, *64*, 222 (1977).

141. H. J. Heinen, U. Giessmann, and F. W. Röllgen, *Org. Mass Spectrom.*, *12*, 710 (1977).

142. H. D. Beckey and F. W. Röllgen, *Org. Mass Spectrom.*, submitted for publication.

143. F. W. Röllgen and H.-R. Schulten, *Z. Naturforsch., 30a*, 1685 (1975).

144. H. J. Veith, *Angew. Chem., 88*, 762 (1976); *Angew. Chem. Int. Ed. Eng., 15*, 695 (1976).

145. J.-C. Prome and G. Puzo, *Org. Mass Spectrom., 12*, 28 (1977).

146. W. D. Lehmann and H.-R. Schulten, *Anal. Chem., 49*, 1744 (1977).

147. H.-R. Schulten, R. Ziskoven, and W. D. Lehmann, *Z. Naturforsch., 33c*, 178 (1978).

148. H.-R. Schulten, W. D. Lehmann, and R. Ziskoven, *Z. Naturforsch., 33c*, 484 (1978).

149. W. D. Lehmann, U. Bahr, and H.-R. Schulten, *Biomed. Mass Spectrom., 5*, 536 (1978).

150. W. D. Lehmann and H.-R. Schulten, *Angew. Chem., 89*, 890 (1977); *Angew. Chem. Int. Ed. Eng., 16*, 852 (1977).

151. H.-R. Schulten and W. D. Lehmann, *Mikrochim. Acta*, (Wien) 1978 *II*, 113.

152. H. D. Lehmann, H.-R. Schulten, and H. M. Schiebel, *Z. Anal. Chem., 289*, 11 (1978).

153. W. D. Lehmann and H.-R. Schulten, *Biomed. Mass Spectrom., 5*, 208 (1978).

154. H.-R. Schulten and W. D. Lehmann, in *Quantitative Mass Spectrometry in Life Sciences II* (A. P. De Leenheer, R. Roncucci, and C. Van Peteghem, Eds.), Elsevier Scientific Publ., Amsterdam, 1978.

155. N. Evans, D. E. Games, A. H. Jackson, and S. A. Matlin, *J. Chromatogr., 115*, 325 (1975).

156. J. F. J. Hughes, N. Evans, D. E. Games, M. J. E. Hewlins, A. H. Jackson, J. R. Jackson, N. A. Khan, S. A. Matlin, M. Rossiter, R. G. Saxton, H. A. Swaine, and K. T. Taylor, in *Advances in Mass Spectrometry in Biochemistry and Medicine, Vol. 1* (A. Frigerio and N. Gastagnoli, Eds.), Spectrum Publications, New York, 357 (1976).

157. C. N. McEwen, S. F. Layton, and S. K. Taylor, *Anal. Chem., 49*, 922 (1977).

158. R. H. Wiley and J. Carter Cook, Jr., *J. Macromol. Sci.-Chem., A10*, 811 (1976).

159. K. L. Olson, K. L. Rinehart, Jr., and J. Carter Cook, Jr., *Biomed. Mass Spectrom., 4*, 284 (1977).

160. R. C. Pandey, J. Carter Cook, Jr., and K. L. Rinehart, Jr. *J. Am. Chem. Soc., 99*, 8469 (1977).

161. H.-R. Schulten, T. Komori, and T. Kawasaki, *Tetrahedron, 33*, 2595 (1977).

162. H.-R. Schulten, T. Komori, T. Nohara, R. Higuchi, and T. Kawasaki, *Tetrahedron, 34*, 1003 (1978).

163. H.-R. Schulten, W. D. Lehmann, and D. Haaks, *Org. Mass Spectrom., 13*, 361 (1978).

164. H.-R. Schulten and H. M. Schiebel, *Naturwiss., 65*, 223 (1978).

AUTHOR INDEX

Numbers in parentheses are reference numbers and indicate that an author's work is referred to although his name is not cited in the text. Underlined numbers give the page on which the complete reference is listed.

In Chapter 5, References 137-164 were added in proof (pp. 264-266) and are not included in the index.

C

Y

Z

SUBJECT INDEX